WiMAX

WiMAX
TECHNOLOGY FOR BROADBAND WIRELESS ACCESS

Loutfi Nuaymi

ENST Bretagne,

France

John Wiley & Sons, Ltd

Other Wiley Editorial Offices

John Wiley & Sons Inc., 111 River Street, Hoboken, NJ 07030, USA

Jossey-Bass, 989 Market Street, San Francisco, CA 94103-1741, USA

Wiley-VCH Verlag GmbH, Boschstr. 12, D-69469 Weinheim, Germany

John Wiley & Sons Australia Ltd, 42 McDougall Street, Milton, Queensland 4064, Australia

John Wiley & Sons (Asia) Pte Ltd, 2 Clementi Loop #02-01, Jin Xing Distripark, Singapore 129809

John Wiley & Sons Canada Ltd, 6045 Freemont Blvd, Mississauga, ONT, L5R 4J3

Wiley also publishes its books in a variety of electronic formats. Some content that appears in print may not be
available in electronic books.

This book contains text excerpts, tables and figures reprinted with permission from IEEE Std 802.16 [IEEE
802.16-2004, IEEE Standard for Local and Metropolitan Area Networks, Air Interface for Fixed Broadband
Wireless Access Systems, Oct. 2004; IEEE 802.16f, Amendment 1: Management Information Base, Dec. 2005;
IEEE 802.16e, Amendment 2: Physical and Medium Access Control Layers for Combined Fixed and Mobile
Operation in Licensed Bands and Corrigendum 1, Feb. 2006], Copyright IEEE 2007, by IEEE. The IEEE
disclaims any responsibility or liability resulting from the placement and use in the described manner.

British Library Cataloguing in Publication Data

A catalogue record for this book is available from the British Library

ISBN 978-0-470-02808-7 (HB)

Typeset in 10/12 pt Times Roman by Thomson Digital.
Printed and bound in Great Britain by Antony Rowe Ltd, Chippenham, England.
This book is printed on acid-free paper responsibly manufactured from sustainable forestry
in which at least two trees are planted for each one used for paper production.

To my wife, Gaëlle,

and our lovely daughter,

Alice

Contents

Preface and Acknowledgements

WiMAX technology is presently one of the most promising global telecommunication systems. Great hopes and important investments have been made for WiMAX, which is a Broadband Wireless Access System having many applications: fixed or last-mile wireless access, backhauling, mobile cellular network, telemetering, etc. WiMAX is based on the IEEE 802.16 standard, having a rich set of features. This standard defines the Medium Access Layer and the Physical Layer of a fixed and mobile Broadband Wireless Access System. WiMAX is also based on the WiMAX Forum documents.

This book is intended to be a complete introduction to the WiMAX System without having the ambition to replace thousands of pages of documents of the IEEE 802.16 standard and amendments and WiMAX Forum documents. There will always be a need to refer to these for any technical development of a specific aspect of WiMAX.

Besides my teaching of other wireless systems (GSM/GPRS, UMTS and WiFi) and related research, I had the occasion to write a first presentation about WiMAX technology, by coincidence, in 2003 and then a WiMAX report. Student projects, PhD work and wireless network courses teaching then provided me with the building blocks for a first WiMAX document. Starting from February 2006, providing ENST Bretagne Inter-Enterprise training and WiMAX training for other specific companies allowed me to develop an even more complete presentation of WiMAX, using text and slides. I thought it might be helpful for colleague engineers, IT managers and undergraduate and graduate students to use this document as a clear and complete introduction to WiMAX technology. WiMAX users can then, if needed, access more easily some specific part of the standard for a specific development.

Some repetitions will be found in this book. This has been done on purpose in order to provide a complete description of the different aspects of this powerful but also sometimes complex technology.

The book can be divided globally into four independent parts. Part I, Chapters 1 to 4, is a global introduction to WiMAX. Part II, Chapters 5 and 6, describes the physical layer with a focus on the main features of the WiMAX physical layer, OFDM transmission and its OFDMA variant. Part III, Chapters 7 to 11, describes the MAC layer and, more specifically, the multiple access and the QoS Management of WiMAX. Part IV, Chapters 12 to 16, covers diverse topics: radio resource management, the network architecture, mobility and security. The book ends with some comparisons and a conclusion.

Without doubt, this book about such a recent technology could not have been published so early without precious help. I wish to thank Jérôme Brouet, from Alcatel, who agreed to write large parts of Chapters 12 and 13. His excellent knowledge of WiMAX has always been a great help to me. I thank trainee student Gérard Assaf for the very good work he provided for figures, synthesis notes and bibliography notes. I also thank trainee students and ENST Bretagne students Aymen Belghith, Maël Boutin, Matthieu Jubin, Ziad Noun and Badih Souhaid for the same type of help. Other student reports and projects were also useful.

I am grateful for the discussions and comments of (the list is not exhaustive) Olfa Ben Haddada, Luc Brignol, Nora Cuppens, Guillaume Lebrun, Bertrand Léonard and Bruno Tharon and my colleagues Xavier Lagrange, Laurence Rouillé and Philippe Godlewski. The wide knowledge of Francis Dupont about Internet and network security (and, by the way, a lot of other topics) helped me with the security chapter. Walid Hachem provided precious help. My colleague Xavier Lagrange provided total support for this book project.

I also wish to thank Prakash Iyer and Bruce Holloway from the WiMAX Forum for precious remarks and authorisations.

I acknowledge the reason for the existence of this book, the IEEE 802.16-2004 standard and its amendment 802.16e and WiMAX Forum Documents. I wish to thank the authors of these documents.

Sarah Hinton, my Project Editor at John Wiley & Sons, Ltd was extremely patient with me. In addition, she helped me a lot with this project.

I thank my parents-in-law Michelle and Marcel for their total support during the marathon last sprint when I invaded Marcel's office for three complete weeks, day and night. My mother Neema also had her share of this book effort.

I end these acknowledgements with the most important: I thank Gaëlle for her support throughout the long writing times. Our little wonder Alice provided me with some of the charming energy she spent for her first steps while I was finishing the book.

I did my best to produce an error-free book and to mention the source of every piece of information. I welcome any comment or suggestion for improvements or changes that could be implemented in possible future editions of this book. The email address for gathering feedback is wimax-thebook@mlistes.enst-bretagne.fr.

Abbreviations List

This list contains the main abbreviations used throughout this book. First the general list is given and then the QoS Classes, the MAC management messages and the security abbreviations list.

3G	Third-generation cellular system. Examples: UMTS and cdma2000
AAA	Authentication Authorisation and Accounting. Protocol realising these three functions. Often related to an AAA server
AAS	Adaptive Antenna Systems. The WiMAX MAC Layer has functionalities that allow the use of AAS
ACK	ACKnowledge or ACKnowledgement. Control message used in the ARQ mechanism
AMC	Adaptive Modulation and Coding
ARCEP	(French telecommunications regulation authority) Autorité de Régulation des Communications Electroniques et des Postes. Old name: ART (Autorité de Régulation des Télécommunications)
ARQ	Automatic Repeat reQuest. Layer two transmission protocol
ASN	Access Service Network. The WiMAX radio access network, mainly composed of BSs and ASN-GW
ASN-GW	ASN Gateway. ASN equipment, between BSs and CSN
ASP	Application Service Provider. Business entity that provides applications or services via (Visited) V-NSP or (Home) H-NSP
ATM	Asynchronous Transfer Mode
BE	Best Effort. BE is one of the five QoS classes of WiMAX. Used for lowest priority time-constraint services such as email
BER	Bit Error Rate
BF	Beamforming. Adaptive Antenna Systems technology
BPSK	Binary Phase Shift Keying. Binary digital modulation
BR	Bandwidth Request
BS	Base Station
BSID	Base Station IDentifier
BSN	Block Sequence Number. Used in Selective ACK variant of the ARQ mechanism
BTC	Block Turbo Code. Turbo coding variant
BW	Bandwidth
BWA	Broadband Wireless Access. High data rate radio access. WiMAX is a BWA
CALEA	Communications Assistance Law Enforcement Act
CBR	Constant Bit Rate. Data transmission service type (e.g. non-optimised voice)

CC	Convolution Code
CDMA	Code Division Multiple Access
CID	Connection IDentifier. A 16-bit identification of a MAC connection
CINR	Carrier-to-Interference-and-Noise Ratio. Also known as the SNR (Signal-to-Noise Ratio)
CLEC	Competitive Local Exchange Carrier. New Operator
CP	Cyclic Prefix. See OFDM theory
CPE	Consumer Premises Equipment. User equipment
CPS	Common Part Sublayer. Middle part of the IEEE 802.16 MAC Layer
CQI	Channel Quality Information. A CQI is transmitted on a CQI channel
CQICH	Channel Quality Information CHannel. The BS may allocate a CQICH subchannel for channel state information fast-feedback
CRC	Cyclic Redundancy Check
CS	Convergence Sublayer. Higher part of the IEEE 802.16 MAC Layer. The Service-Specific Convergence Sublayer (CS) realises the transformation and/or the mapping of external network data before its transmission on a 802.16 radio link
CSN	Connectivity Service Network (cf. Architecture WiMAX). Set of network functions that provide IP connectivity services to the WiMAX subscriber(s). A CSN may comprise network elements such as routers, AAA proxy/servers, user databases and interworking gateway devices
CT2/CAI	Cordless Telephone 2 / Common Air Interface. Digital WLL cordless phone system
CTC	Convolutional Turbo Code. Turbo coding variant
DAMA	Demand Assigned Multiple Access
DC	Direct Current
DCD	Downlink Channel Descriptor. Downlink Descriptor MAC Management message
DECT	Digital Enhanced Cordless Telecommunications. Cordless phone system
DFS	Dynamic Frequency Selection
DHCP	Dynamic Host Configuration Protocol. The DHCP server provides the DHCP client with configuration informations, in particular, an IP address
DIUC	Downlink Interval Usage Code. Burst profile identifier, accompanying each downlink burst
DL	DownLink
DLFP	DownLink Frame Prefix. Position and burst profile of the first downlink burst are provided in DLFP. DLFP is in FCH
DL-MAP	DownLink MAP. MAC Management message, transmitted at the beginning of a downlink frame, indicating its contents
DNS	Domain Name System
DSL	Digital Subscriber Line
EC	Encryption Control. Generic Header bit
EIRP	Equivalent Isotropic Radiated Power
EKS	Encryption Key Sequence. Generic Header field
ertPS	Extended real-time Polling Service. New QoS class added by the 802.16e amendment

FA	Foreign Agent
FBSS	Fast BS Switching. Fast make-before-break handover
FCH	Frame Control Header. Downlink frame header
FDD	Frequency Division Duplexing
FEC	Forward Error Correction. Channel coding
FFT	Fast Fourier Transform. Matrix computation that allows the discrete Fourier transform to be computed (while respecting certain conditions)
FSN	Fragment Sequence Number
FTP	File Transfer Protocol
FUSC	Full Usage of the SubChannels. OFDMA Permutation mode
GMH	Generic MAC Header
GSM	Global System for Mobile communication. Second-generation cellular system
HA	Home Agent
HARQ	Hybrid Automatic Repeat reQuest. Evolution of ARQ protocol. Sometimes denoted H-ARQ
HCS	Header Check Sequence
H-FDD	Half-duplex FDD
HLR	Home Location Register
H-NSP	Home NSP
HO	HandOver
HT	Header Type. MAC header bit
HUMAN	High-speed Unlicensed Metropolitan Area Network. Free license 802.16 specification
IE	Information Element. Element of a MAC message. For example, a DL-MAP_ IE describes one burst profile
IEEE	Institute of Electrical and Electronics Engineers
IETF	Internet Engineering Task Force
IFFT	Inverse Fast Fourier Transform. OFDM theory shows that an IFFT operation application leads to orthogonal frequencies (also called subcarriers or tones)
ILEC	Incumbent Local Exchange Carrier
IMS	IP Multimedia Subsystem
IP	Internet Protocol
ISM	Industrial, Scientific and Medical. Appellation of the unlicensed 2.4 GHz frequency bandwidth
IUC	Interval Usage Code. See DIUC and UIUC
LDPC	Low-Density Parity Check code. Channel coding
LEN	LENgth. Length in bytes of a MAC PDU. Includes the MAC header and, if present, the CRC
LoS	Line-of-Sight. A radio transmission is LoS if it fulfills certain conditions (Fresnel zone sufficiently clear)
LTE	Long-Term Evolution. Evolution of the 3G system
MAC	Media Access Control Layer. Part of Layer 2 of the OSI Networks Model
MAC	Message Authentication Code. The ciphertext Message Authentication Code, also known as MAC, must not be confused with the Medium Access Layer, MAC. Except in Section 15.4, MAC is used for the Medium Access Control Layer

MAN	Metropolitan Area Network. IEEE 802.16 is a Wireless MAN system
MBS	Multicast and Broadcast Services feature
MCS	Modulation and Coding Scheme
MDHO	Macro Diversity HandOver. A state where the mobile communicates with more than one BS
MIB	Management Information Base. The BS and SS managed nodes collect and store the managed objects in an 802.16 MIB format
MIMO	Multiple-Input Multiple-Output
MIP	Mobile IP
MMDS	Multichannel Multipoint Distribution Service
MPDU	MAC PDU
MS	Mobile Station
MSDU	MAC SDU
NACK	Non-ACKnowledge or Non-ACKnowledgement. Control message used in the ARQ mechanism
NAP	Network Access Provider (cf. Architecture WiMAX). Business entity that provides a WiMAX radio access infrastructure to one or more WiMAX Network Services
NLoS	Non-Line-of-Sight. A radio transmission is NLoS if it do not fulfil certain conditions (Fresnel zone sufficiently clear)
nrtPS	Non-real-time Polling Services. One of the five QoS classes of WiMAX
NSP	Network Service Provider (cf. Architecture WiMAX). Business entity that provides IP connectivity and WiMAX services to WiMAX subscribers
NWG	NetWork Group. WiMAX Forum Group. In charge of creating the high-level architecture specifications
OEM	Original Equipment Manufacturer
OFDM	Orthogonal Frequency Division Multiplexing. Transmission technique. The principle is to transmit the information on many orthogonal frequency subcarriers
OFDMA	Orthogonal Frequency Division Multiple Access. OFDM used as a multiple access scheme
OPUSC	Optional PUSC
PAPR	Peak-to-Average Power Ratio. In an OFDM transmission, the PAPR is the peak value of transmitted subcarriers to the average transmitted signal
PBR	PiggyBack Request. Grant Management subheader field indicating the uplink bandwidth requested by the SS
PCM	Pulse Coded Modulation. Classical phone signal transmission system. Variants are T1 and E1
PDU	Protocol Data Unit
PHS	Payload Header Suppression. Optional CS sublayer process
PHSF	Payload Header Suppression Field
PHSI	Payload Header Suppression Index
PHSM	Payload Header Suppression Mask
PHSS	Payload Header Suppression Size
PHSV	Payload Header Suppression Valid
PHY	PHYsical layer

PICS	Protocol Implementation Conformance Specification document. In the conformance test, the BS/SS units must pass all mandatory and prohibited test conditions called out by the test plan for a specific system profile.
PM	Poll-Me bit. SSs with currently active UGS connections may set the PM bit (in the Grant Management subheader) in a MAC packet of the UGS connection to indicate to the BS that they need to be polled to request bandwidth for non-UGS connections
PMP	Point-to-MultiPoint. Basic WiMAX topology
PN	Pseudo-Noise sequence
PRBS	Pseudo-Random Binary Sequence. Used in the randomisation block
PS	Physical Slot. Function of the PHYsical Layer. Used as a resource attribution unit
PUSC	Partial Usage of SubChannels. OFDMA Permutation mode
QAM	Quadrature Amplitude Modulation
QoS	Quality of Service
QPSK	Quadrature Phase Shift Keying
RF	Radio Frequency
RFC	Request For Comment. IETF document
RRA	Radio Resource Agent
RRC	Radio Resource Controller
RRM	Radio Resource Management
RS	Reed–Solomon code. Channel coding
RSSI	Received Signal Strength Indicator. Indicator of the signal-received power level
RTG	Receive/transmit Transition Gap. The RTG is a gap between the uplink burst and the subsequent downlink burst in a TDD transceiver
RTP	Real-Time Protocol
rtPS	Real-time Polling Services. One of the five QoS classes of WiMAX
SAP	Service Access Point
SBC	SS Basic Capability. The BS and the SS agree on the SBC at SS network entry
SC	Single Carrier. A single carrier transmission is a transmission where no OFDM is applied
SDU	Service Data Unit
SFA	Service Flow Authorisation
SFID	Service Flow IDentifier. An MAC service flow is identified by a 32-bit SFID
SFM	Service Flow Management
SI	Slip Indicator. Grant Management subheader field. Indicates slip of uplink grants relative to the uplink queue depth
SISO	Single-Input Single-Output. Specific case of MIMO
SLA	Service Level Agreements
SM	Spatial Multiplexing. MIMO family of algorithms
SN	Sequence Number. Transmitted block number used in the ARQ mechanism
SNMP	Simple Network Management Protocol. IETF Network Management Reference model protocol
SNR	Signal-to-Noise Ratio. The noise includes interferer signals. Also known as CINR (Carrier-to-Interference-and-Noise Ratio)

SOFDMA	Scalable OFDMA
SPID	SubPacket IDentifier. Used in the HARQ process
SS	Subscriber Station
STBC	Space Time Block Coding. MIMO variant
STC	Space Time Coding. MIMO variant
TCP	Transmission Control Protocol
TCS	Transmission Convergence Sublayer. Optional PHY mechanism
TDD	Time Division Duplexing
TDM	Time Division Multiplexing. A TDM burst is a contiguous portion of a TDM data stream using the same PHY parameters. These parameters remain constant for the duration of the burst. TDM bursts are not separated by gaps or preambles
TFTP	Trivial File Transfer Protocol
TLV	Type/Length/Value
TO	Transmission Opportunity
TTG	Tx/Rx Transition Gap. Time gap between the downlink burst and the subsequent uplink burst in the TDD mode
TUSC	Tile Usage of SubChannels. OFDMA Permutation mode. Two variants: TUSC1 and TUSC2
UDP	User Datagram Protocol
UDR	Usage Data Records
UCD	Uplink Channel Descriptor. Uplink Descriptor MAC Management message
UGS	Unsolicited Grant Services. One of the five QoS classes of WiMAX
UIUC	Uplink Interval Usage Code. Burst profile identifier, accompanying each uplink burst
UL	UpLink
UL-MAP	UpLink MAP. The MAC Management message indicating the contents of an uplink frame
UTC	Universal Coordinated Time
V-NSP	Visited NSP
VoIP	Voice over IP
WiFi	Wireless Fidelity. IEEE 802.11 certification consortium
WiMAX	Worldwide Interoperability for Microwave Access Forum. The WiMAX Forum provides certification of conformity, compatibility and interoperability of IEEE 802.16 products. In extension WiMAX is also the common name for the technology mainly based on IEEE 802.16
WLL	Wireless Local Loop. Cordless phone system

IEEE 802.16 Qos Classes (or Service Classes)

BE	Best Effort. Used for lowest priority time-constraint services such as email
ertPS	Extended real-time Polling Service. New QoS class defined in the 802.16e amendment. Intermediary between rtPS and UGS
nrtPS	Non-real-time Polling Services. Used for non-real-time services having some time constraints

rtPS   Real-time Polling Services. Used for variable data rate real-time services. Example is the MPEG video

UGS   Unsolicited Grant Services. Dedicated to Constant Bit Rate (CBR) services, UGS guarantees fixed-size data packets issued at periodic intervals. Example of use is T1/E1 transmissions

IEEE 802.16 MAC Management Messages

Note that more details of the MAC Management messages can be found in Annex A.

AAS-BEAM_REQ	AAS Beam REQuest message
AAS-BEAM_RSP	AAS Beam ReSPonse message
AAS-Beam_Select	AAS Beam Select message
AAS-FBCK-REQ	AAS FeedBaCK REQuest message
AAS-FBCK-RSP	AAS FeedBaCK ReSPonse message
ARQ-Discard	ARQ Discard message
ARQ-Feedback	Standalone ARQ Feedback message
ARQ- Reset	ARQ Reset message
CLK-CMP	SS network CLocK CoMParison message
DBPC-REQ	Downlink Burst Profile Change REQuest message
DBPC-RSP	Downlink Burst Profile Change ReSPonse message
DCD	Downlink Channel Descriptor message
DL-MAP	DownLink Access Definition message
DREG-CMD	De/re-REGister CoMmanD message
DREG-REQ	SS De-REGistration message
DSA-ACK	Dynamic Service Addition ACKnowledge message
DSA-REQ	Dynamic Service Addition REQuest message
DSA-RSP	Dynamic Service Addition ReSPonse message
DSC-ACK	Dynamic Service Addition ACKnowledge message
DSC-REQ	Dynamic Service Change REQuest message
DSC-RSP	Dynamic Service Change ReSPonse message
DSD-REQ	Dynamic Service Deletion REQuest message
DSD-RSP	Dynamic Service Deletion ReSPonse message
DSX-RVD	DSx ReceiVeD message
FPC	Fast Power Control message
MBS_MAP	MBS MAP message
MCA-REQ	MultiCast Assignment REQuest message
MCA-RSP	MultiCast Assignment ReSPonse message
MSH-CSCF	MeSH Centralised Schedule ConFiguration message
MSH-CSCH	MeSH Centralised SCHedule message
MSH-DSCH	MeSH Distributed SCHedule message
MSH-NCFG	MeSH Network ConFiGuration message
MSH-NENT	MeSH Network ENTry message
MOB_ASC-REP	ASsoCiation result REPort message
MOB_BSHO-REQ	BS HO REQuest message

MOB_BSHO-RSP	BS HO ReSPonse message
MOB_HO-IND	HO INDication message
MOB_MSHO-REQ	MS HO REQuest message
MOB_NBR-ADV	NeighBouR ADVertisement message
MOB_PAG-ADV	BS broadcast PAGing Advertisement message
MOB_SCN-REQ	SCaNning interval allocation REQuest message
MOB_SCN-RSP	SCaNning interval allocation ReSPonse message
MOB_SCN-REP	SCaNning result REPort message
MOB_SLP-REQ	SLeeP REQuest message
MOB_SLP-RSP	SLeeP ReSPonse message
MOB_TRF-IND	TRaFfic INDication message
PKM-REQ	Privacy Key Management REQuest message
PKM-RSP	Privacy Key Management ReSPonse message
PMC_REQ	Power control Mode Change REQuest message
PMC_RSP	Power control Mode Change ReSPonse message
PRC-LT-CTRL	Setup/tear-down of Long-Term MIMO precoding message
REG-REQ	REGistration REQuest message
REG-RSP	REGistration ReSPonse message
REP-REQ	Channel measurement REPort REQuest message
REP-RSP	Channel measurement REPort ReSPonse message
RES-CMD	RESet CoMmanD message
RNG-REQ	RaNGing REQuest message
RNG-RSP	RaNGing ReSPonse message
SBC-REQ	SS Basic Capability REQuest message
SBC-RSP	SS Basic Capability ReSPonse message
TFTP-CPLT	Config File TFTP ComPLeTe Message
TFTP-RSP	Config File TFTP complete ReSPonse message
UCD	Uplink Channel Descriptor message
UL-MAP	UpLink Access Definition message

Security Abbreviations

AES	Advanced Encryption Standard. The AES Algorithm is a shared (secret)-key encryption algorithm
AK	Authorisation Key (PKMv1 and PKMv2)
CA	Certification Authority
CBC	Cipher Block Chaining mode. An AES mode
CCM	Counter with CBC-MAC (CBC: Cipher Block Chaining mode). AES CCM is an authenticate-and-encrypt block cipher mode used in IEEE 802.16 for data encryption
CMAC	Cipher-based Message Authentication Code
CMAC_KEY_D	CMAC KEY for the Downlink. Used for authenticating messages in the downlink direction
CMAC_KEY_U	CMAC KEY for the Uplink. Used for authenticating messages in the uplink direction
DES	Data Encryption Standard. Shared (secret)-key encryption algorithm

EAP	Extensible Authentication Protocol. Mutual authentification protocol framework
EIK	EAP Integrity Key
GKEK	Group Key Encryption Key (PKMv2)
GTEK	Group Traffic Encryption Key (PKMv2)
HMAC	Hashed Message Authentication Code
HMAC_KEY_D	HMAC Key for the Downlink. Used for authenticating messages in the downlink direction
HMAC_KEY_S	HMAC Key in the Mesh mode
HMAC_KEY_U	HMAC Key for the Uplink. Used for authenticating messages in the uplink direction
KEK	Key Encryption Key (PKMv1 and PKMv2)
MAK	MBS Authorisation Key (PKMv2)
MGTEK	MBS Group Traffic Encryption Key (PKMv2)
MTK	MBS Traffic Key (PKMv2)
PAK	Primary Authorisation Key (PKMv2)
PKM	Privacy Key Management protocol
PMK	Pairwise Master Key (PKMv2)
PN	Packet Number
RSA	Rivest Shamir Adleman. Public key encryption algorithm used to encrypt some MAC management security messages, using the SS public key
SA	Security Association. Set of security information agreed between a BS and one or more of its client SSs (methods for data encryption, data authentication, keys exchange, etc.)
SAID	Security Association IDentifier. A 16-bit identifier shared between the BS and the SS that uniquely identifies a security association
SHA	Secure Hash algorithm
TEK	Traffic Encryption Key (PKMv1 and PKMv2)

Part One

Global Introduction to WiMAX

1

Introduction to Broadband Wireless Access

1.1 The Need for Wireless Data Transmission

Since the final decades of the twentieth century, data networks have known steadily growing success. After the installation of fixed Internet networks in many places all over the planet and their now large expansion, the need is now becoming more important for wireless access. There is no doubt that by the end of the first decade of the twentieth century, high-speed wireless data access, i.e. in Mb/s, will be largely deployed worldwide.

Wireless communication dates back to the end of the nineteenth century when the Maxwell equations showed that the transmission of information could be achieved without the need for a wire. A few years later, experimentations such as those of Marconi proved that wireless transmission may be a reality and for rather long distances. Through the twentieth century, great electronic and propagation discoveries and inventions gave way to many wireless transmission systems.

In the 1970s, the Bell Labs proposed the cellular concept, a magic idea that allowed the coverage of a zone as large as needed using a fixed frequency bandwidth. Since then, many wireless technologies had large utilisation, the most successful until now being GSM, the Global System for Mobile communication (previously Groupe Spécial Mobile), originally European second generation cellular system. GSM is a technology mainly used for voice transmission in addition to low-speed data transmission such as the Short Message Service (SMS).

The GSM has evolutions that are already used in many countries. These evolutions are destined to facilitate relatively high-speed data communication in GSM-based networks. The most important evolutions are:

- GPRS (General Packet Radio Service), the packet-switched evolution of GSM;
- EDGE (Enhanced Data rates for GSM Evolution), which includes link or digital modulation efficiency adaptation, i.e. adaptation of transmission properties to the (quickly varying) radio channel state.

In addition to GSM, third-generation (3G) cellular systems, originally European and Japanese UMTS (Universal Mobile Telecommunication System) technology and originally American cdma2000 technology, are already deployed and are promising wireless communication systems.

WiMAX: Technology for Broadband Wireless Access   Loutfi Nuaymi
© 2007 John Wiley & Sons, Ltd

Cellular systems have to cover wide areas, as large as countries. Another approach is to use wireless access networks, which were initially proposed for Local Area Networks (LANs) but can also be used for wide area networks.

1.2 Wireless Networks and Broadband Wireless Access (BWA)

1.2.1 Different Types of Data Networks

A large number of wireless transmission technologies exist, other systems still being under design. These technologies can be distributed over different network families, based on a network scale. In Figure 1.1, a now-classical representation (sometimes called the 'eggs figure') is shown of wireless network categories, with the most famous technologies for each type of network.

A *Personal Area Network* (PAN) is a (generally wireless) data network used for communication among data devices close to one person. The scope of a PAN is then of the order of a few metres, generally assumed to be less than 10 m, although some WPAN technologies may have a greater reach. Examples of WPAN technologies are Bluetooth, UWB and Zigbee.

A *Local Area Network* (LAN) is a data network used for communication among data devices: computer, telephones, printer and personal digital assistants (PDAs). This network covers a relatively small area, like a home, an office or a small campus (or part of a campus). The scope of a LAN is of the order of 100 metres. The most (by far) presently used LANs are Ethernet (fixed LAN) and WiFi (Wireless LAN, or WLAN).

A *Metropolitan Area Network* (MAN) is a data network that may cover up to several kilometres, typically a large campus or a city. For instance, a university may have a MAN that joins together many of its LANs situated around the site, each LAN being of the order

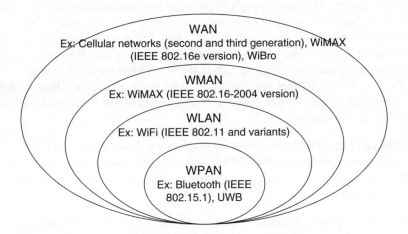

Figure 1.1 Illustration of network types. For each category, the most well known technologies are given. To this figure, some people add a smaller 'egg' in the WPAN (Wireless Personal Area Network), representing the WBAN (Wireless Body Area Network), with a coverage of the magnitude of a few metres, i.e. the proximity of a given person

of half a square kilometre. Then from this MAN the university could have several links to other MANs that make up a WAN. Examples of MAN technologies are FDDI (Fiber-Distributed Data Interface), DQDB (Distributed Queue Dual Bus) and Ethernet-based MAN. Fixed WiMAX can be considered as a Wireless MAN (WMAN).

A *Wide Area Network* (WAN) is a data network covering a wide geographical area, as big as the Planet. WANs are based on the connection of LANs, allowing users in one location to communicate with users in other locations. Typically, a WAN consists of a number of interconnected switching nodes. These connections are made using leased lines and circuit-switched and packet-switched methods. The most (by far) presently used WAN is the Internet network. Other examples are 3G and mobile WiMAX networks, which are Wireless WANs. The WANs often have much smaller data rates than LANs (consider, for example, the Internet and Ethernet).

To this figure, some people add a smaller 'egg' in the WPAN, representing the WBAN, Wireless Body Area Network, with a coverage of the magnitude of a few metres, i.e. the near proximity of a given person. A WBAN may connect, for example, the handset to the ear-phone, to the 'intelligent' cloth, etc.

1.2.2 Some IEEE 802 Data Network Standards

WiMAX is based on the IEEE 802.16 standard [1,2]. Standardisation efforts for local area data networks started in 1979 in the IEEE, the Institute of Electrical and Electronics Engineers. In February 1980 (80/2), the IEEE 802 working group (or committee) was founded, dedicated to the definition of IEEE standards for LANs and MANs. The protocols and services specified in IEEE 802 map to the lower two layers (Data Link and Physical) of the seven-layer OSI networking reference model [3,4]. IEEE 802 splits the OSI Data Link Layer into two sublayers named Logical Link Control (LLC) and Media Access Control (MAC) (see Chapter 3).

Many subcommittees of IEEE 802 have since been created. The most widely used network technologies based on IEEE 802 subcommittees are the following:

- IEEE 802.2, Logical Link Control (LLC). The LLC sublayer presents a uniform interface to the user of the data link service, usually the network layer (Layer 3 of the OSI model).
- IEEE 802.3, Ethernet. The Ethernet, standardised by IEEE 802.3, is a family of network technologies for LANs, standardized by IEEE 802.3. It quickly became the most widespread LAN technology until the present time. Possible data rates are 100 Mb/s, 1 Gb/s and 10 Gb/s.
- IEEE 802.5, Token Ring. The Token Ring LAN technology was promoted by IBM in the early 1980s and standardised by IEEE 802.5. Initially rather successful, Token Ring lost ground after the introduction of the 10BASE-T evolution of Ethernet in the 1990s.
- IEEE 802.11, WLAN. IEEE 802.11 is the subcommittee that created what is now known as WiFi Technology. A Wireless Local Area Network (WLAN) system and many variants were proposed by the IEEE 802.11 working group (and subcommittees), founded in 1990. A WLAN covers an area whose radius is of the magnitude of 100 metres (300 feet). First, IEEE 802.11 (www.ieee802.org/11/) and its two physical radio link variants, 802.11a and 802.11b standards, were proposed by the end of the 1990s. IEEE 802.11b products, certified by WiFi (Wireless Fidelity) Consortium, were available soon after. These products have nearly always been known as being of WiFi Technology. These WiFi products

quickly encountered a large success, mainly due to their simplicity but also the robustness
of the technology, in addition to the relative low cost and the use of unlicensed 2.4 GHz and
5 GHz frequency bands. Other variants of the basic 802.11 standard are available (802.11e,
802.11g, 802.11h, 802.11i, etc.) or are at the draft stage (802.11n, etc.).

- IEEE 802.15, WPAN. Different WPAN technologies were or are defined in IEEE 802.15.
  IEEE 802.15.1 included Bluetooth, initially proposed by a consortium of manufacturers,
  and now studies the evolution of Bluetooth. Bluetooth is now a widely used (data) cable-
  replacement technology with a theoretical scope of up to 20 m. IEEE 802.15.3a studied an
  Ultra-Wide Band (UWB) System, very high-speed and very low-distance network. The
  IEEE 802.15.3a draft has not yet been approved. IEEE 802.15.4 is about ZigBee, a low-
  complexity technology for automatic application and an industrial environment.
- IEEE 802.16, BWA. IEEE 802.16 is the working group of IEEE 802 dedicated to BWA.
  Its aim is to propose standards for (high data rate) WMAN. IEEE 802.16 standards are
  detailed in Section 2.2. As for 802.11 products a certification forum was created for IEEE
  802.16 products, the WiMAX (Worldwide Interoperability for Microwave Access) forum,
  also described in Chapter 2. It can already be said that WiMAX is the name normally used
  for IEEE 802.16 products.

BWA networks have a much greater range than WLAN WiFi. In fact, IEEE 802.16 BWA has
two variants: IEEE 802.16-2004, which defines a fixed wireless access WMAN technology,
and IEEE 802.16e, which is an amendment of 802.16-2004 approved in December 2005. It
included mobility and then fast handover, then becoming a Wireless WAN (see Figure 1.1).

- IEEE 802.20, Mobile Broadband Wireless Access (MBWA). The aim of this group is to
  define a technology for a packet-based air interface designed for IP (Internet Protocol)-
  based services. This technology is destined for high-speed mobile devices. It was reported
  that MBWA will be based on the so-called Flash OFDM technology proposed by Flarion
  Company.
     A draft 802.20 specification was balloted and approved on 18 January 2006. On 8 June
  2006, the IEEE Standards Board directed that all activities of the 802.20 working group
  be temporarily suspended [3].
- IEEE 802.21, Media Independent Handover (MIH). IEEE 802.21 is a new IEEE standard. It
  is definitely interesting for a telecommunication equipment to have the possibility of realising a handover between two different wireless technologies. A handover is the operation of
  changing the corresponding base station (the cell), the communication channel, the technology, etc., without interruption of an ongoing telecommunication session (conversation
  or other). IEEE 802.21 studies standards enabling handover and interoperability between
  different network types, which is called MIH. These network types can be of the IEEE 802
  family or not. For example, the 802.21 standard would provide information to allow a handover between 3G and 802.11/WiFi networks.

1.2.3 Cordless WLL Phone Systems

Along with progress in cellular (or mobile) systems and wireless data networks, wireless phone systems have began to appear. An important budget for a phone operator or
carrier has always been the local loop, also called the 'last mile', which connects the phone

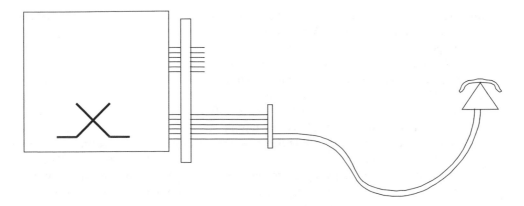

Figure 1.2   Local loop of a classical (voice) phone system

subscriber to the network last elements. It was seen for some configurations that a (radio) Wireless Local Loop (WLL) can be an interesting replacement solution for a fixed (mainly copper) local loop. These WLL systems had to provide a communication circuit, initially for voice, and some low-rate data services. The general principle of a local loop is shown in Figure 1.2.

In a WLL system, terminal stations are connected to a Base Station (BS) through the radio channel (see Figure 1.3). The main difference between WLL and cellular systems is the fact that in a cellular system a subscriber can be connected to one BS or another. A subscriber can also change the BS during a communication without causing an interruption, which is called the handover (or also handoff) procedure.

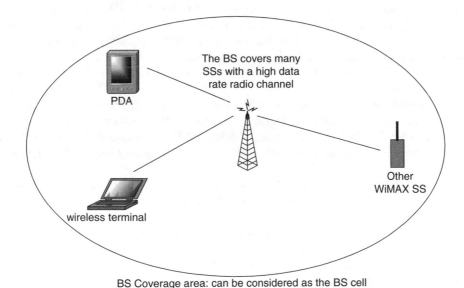

Figure 1.3   Coverage of a given zone by a BS

Several technologies have been proposed for WLL systems, also known as cordless phone systems (or also cordless systems). After analogue systems, mainly proprietary, a digital system was proposed, CT2/CAI (Cordless Telephone 2/Common Air Interface), in 1991. With CT2/CAI, the occupation of one (voice) user is 100 kHz.

The European Telecommunications Standards Institute (ETSI) published a WLL cordless system in 1992 named DECT (Digital Enhanced Cordless Telecommunications). The range of DECT equipments is up to a few hundred metres. DECT works in the 1.9 GHz bandwidth.

DECT is a digital TDMA (Time Division Multiple Access) suited for voice and low data rate applications, in the order of tens of kb/s. Some evolutions of DECT, featuring many slots per user, propose higher data rates up to hundreds of kb/s. DECT has a relatively high success rate nowadays, yet it is a capacity-limited system as TDMA-only systems do not use the bandwidth very efficiently (a user taking many slots leaves very few resources for other users). The wide use of WLL systems for phone communications and some other low data rate communications gave way to high data rate BWA systems, introduced in Section 1.2.2 above and described in further detail in the next section.

1.3 Applications of BWA

As already introduced above with IEEE 802.16, a BWA system is a high data rate (of the order of Mb/s) WMAN or WWAN. A BWA system can be seen as an evolution of WLL systems mainly featuring significantly higher data rates. While WLL systems are mainly destined for voice communications and low data rate (i.e. smaller than 50 kb/s), BWAs' are intended to deliver data flows in Mb/s (or a little lower).

The first application of BWA is fixed-position high data rate access. This access can then evidently be used for Internet, TV and other expected high data rate applications such as Video-on-Demand (VoD). It will also surely be used for other applications that are not really apparent yet. In one word, the first target of BWA is to be a wireless DSL (Digital Subscriber Line, originally called the Digital Subscriber Loop) or also a wireless alternative for the cable. Some business analysts consider that this type of BWA application is interesting only in countries and regions having relatively underdeveloped telecommunications infrastructure. Indeed, using WiMAX for the fixed-position wireless Internet in Paris or New York does not seem economically viable.

Another possible use of high data rate access with BWA is WiFi Backhauling. As shown in Figure 1.4, the Internet so-called backbone is linked to a BS which may be in Line-of-Sight (LOS) of another BS. This has a Non-Line-of-Sight (NLOS) coverage of Subscriber Stations (SSs). The distinction between IEEE 802.16 NLOS and LOS technologies will be detailed in Chapter 2.

The SS in Figure 1.4 is a Consumer Premises Equipment (CPE). The CPE is a radio-including equipment that realises the link between the BS and the terminal equipment(s) of the user. After the CPE, the user may install a terminal such as a Personal Computer (PC) or a TV and may also connect a WiFi Access Point and then a WLAN (the BWA then realizing the WiFi network backhauling). Hence the two main applications of fixed BWA are the wireless last-mile for high data rate and (more specifically) WiFi backhauling. As shown in this figure, a wireless terminal can then be fixed (geographically) or not. This may be the case of a laptop connected to the CPE with a WiFi connection (see the figure).

The fixed access is the first use of BWA, the next step being nomadicity (see Section 1.3.1 for the difference between nomadicity and mobility). A first evolution of the SS will be the

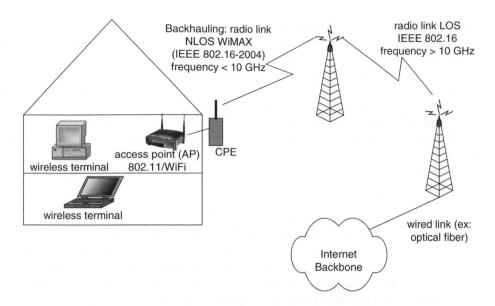

Figure 1.4 Broadband Wireless Access (BWA) applications with a fixed access. The two main applications of a fixed BWA are wireless last-mile for high data rate and (more specifically) WiFi backhauling

case when it is no longer a CPE but a card installed in some laptop. A nomadic access, shown in Figure 1.5, is an access where the user or the subscriber may move in a limited area, e.g. in an apartment or a small campus. This area is the one covered by a BS. Whenever the user moves out of the zone, the communication (or the session) is interrupted. A typical example

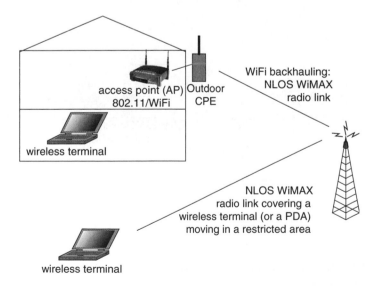

Figure 1.5 Nomadic or portable BWA

of a portable access is WLAN/WiFi use in its first versions (802.11, 802.11b and 802.11a) where a session is interrupted when the terminal gets out of a WLAN coverage even if it enters a zone covered by another WLAN, e.g. in two neighbouring companies.

The nomadic access is very useful in some cases, such as campuses, company areas, compounds, etc. It can be observed that due to this position, which is not fixed, the link between the BS and the SS has to be NLOS (it can be LOS only in the case of fixed CPEs, theoretically). A nomadic access is also sometimes known as a wireless access. The final expected step of WiMAX is a mobile access. The difference between wireless and mobile will now be discussed.

1.3.1 Wireless is Not Mobile!

Different scenarios of mobility can be considered. The most simple one is when two neighbouring BSs belong to the same operator. Hence, the same billing system and customer care apply to the two BSs. In this case, a user moving from one cell to a neighbouring one has to start the session again. This feature is nomadicity rather than mobility. Mobility (or full mobility) is the scenario where the session is not interrupted, whether this is a data session, a voice communication (over IP or not), a video transmission, etc.

The distinction is made between wireless (but yet geographically) fixed access, nomadicity, portability and mobility. Portability is when a user can move with a reasonable speed over a large area, covered by many BSs, without interruption of an possible open session or communication. The value considered as a reasonable speed is of the order of

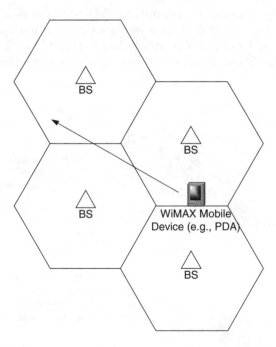

Figure 1.6   Mobile Broadband Wireless Access (BWA). A mobile WiMAX device can move over all the cells in a seamless session

120 km/h. Mobility is the same as portability but with no real limit for speed; i.e. if mobility is realised, a BWA can be used in some high-speed trains with speeds exceeding 350 km/h.

In cellular systems, second generation or later, a voice communication is not interrupted when a mobile moves from one cell to another. This is the so-called 'handover'. The cellular systems are then real mobile networks. Is WiMAX a cellular mobile network? Considering that a cell is the area covered by one BS, the only condition would be a high-speed handover feature. This should be realised with 802.16e evolution of 802.16. However, a WiMAX handover is not expected to occur at very high speeds – to be precise, at speeds higher than a magnitude of 100 km/h. The final objective of WiMAX is to be a mobile system. In this case, part or all of a territory or country will be covered by contiguous cells with a seamless session handover between cells, as in a cellular system (see Figure 1.6). It is evident that WiMAX will then become a rival to 3G cellular systems.

Some service providers define triple play as the combination of data (Internet), voice (unlimited phone calls) and video (TV, video on demand). This evolves into quadruple play by adding mobility. In a first step, this mobility will in fact be only nomadicity, e.g. using the WiMAX subscription to have an Internet access in a café far away from home.

Another application sometimes mentioned for BWA is telemetering: using the BWA for reporting electricity, gas, water, etc. This should represent a small but yet perhaps interesting market. WiMAX telemetering products have already been reported. Evidently, WiMAX is not the only technology that can be used for telemetering.

1.3.2 Synthesis of WiMAX BWA Applications

To sum up, the applications known or expected today of WiMAX as a BWA system are:

- Broadband fixed wireless access. WiMAX would be a competitor for fixed-line high data rate providers in urban and rural environments.
- WiFi backhauling.
- Telemetering. This should represent a small but yet perhaps interesting market.
- Nomadic Internet access.
- Mobile (seamless sessions) high data rate access.

1.4 History of BWA Technologies

1.4.1 Video Distribution: LMDS, MMDS and DVB

The Local Multipoint Distribution Service (LMDS) is a fixed wireless access system specified in the United States by the Digital Audio-Visual Council (Davic), a consortium of video equipment suppliers, network operators and other telecommunication industries. Davic was created in 1993. LMDS is a broadband wireless point-to-multipoint communication technology. Originally designed for wireless digital television transmission, the target applications were then video and Internet in addition to phone.

The standard is rather open and many algorithms used for LMDS are proprietary. Depending on the frequency bandwidth allocated, data rates are of the order of tens of Mb/s in the downlink and Mb/s in the uplink. Link distance can go up to a few km. LMDS operates in the 28 GHz frequency band in the United States. This band is called the LMDS band. Higher frequencies can also be used.

The Multichannel Multipoint Distribution Service (MMDS), also known as wireless cable, is theoretically a BWA technology. It is mainly used as an alternative method of cable television. The MMDS operates on frequencies lower than the LMDS, 2.5 GHz, 2.7 GHz, etc., for lower data rates as channel frequency bandwidths are smaller.

Standardising for digital television started in Europe with the Digital Video Broadcasting (DVB) Project. This standardization was then continued by the European Telecommunications Standard Institute (ETSI). DVB systems distribute data by many mediums: terrestrial television (DVB-T), terrestrial television for handhelds (DVB-H), satellite (DVB-S) and cable (DVB-C). The DVB standards define the physical layer and data link layer of a television distribution system.

Many European countries aim to be fully covered with digital television by around 2010 and to switch off analogue television services by then. DVB will also be used in many places outside Europe, such as India and Australia.

1.4.2 Pre-WiMAX Systems

WiMAX and 802.16 systems will be described in detail in Chapter 2. In this subsection, the pre-WiMAX is introduced. The first version of the IEEE 802.16 standard appeared in 2001. The first complete version was published in 2004. There was evidently a need for wireless broadband much before these dates. Many companies had wireless broadband equipment using proprietary technology since the 1990s and even before. Evidently these products were not interoperable.

With the arrival of the 802.16 standard, many of these products claimed to be based on it. This was again not possible to verify as WiMAX/802.16 interoperability tests and plugfest started in 2006. These products were then known as pre-WiMAX products. Pre-WiMAX equipments were proposed by manufacturers often specialising in broadband wireless. Many of them had important markets in Mexico, Central Europe, China, Lebanon and elsewhere. Device prices were of the order of a few hundred euros. A nonexhaustive list of pre-WiMAX manufacturers contains the following: Airspan, Alvarion, Aperto, Motorola, Navini, NextNet, Proxim, Redline and SR Telecom. Intel and Sequans, among others, provide components.

The performances of pre-WiMAX systems are close to the expected ones of WiMAX, whose products should start to appear from the second part of 2006. Many of the pre-WiMAX equipments were later certified and more are in the process of being certified.

2

WiMAX Genesis and Framework

2.1 IEEE 802.16 Standard

The main features of IEEE 802.16/WiMAX technology are the following:

- (Carrier) frequency <11 GHz. For the moment, the frequency bands considered are 2.5 GHz, 3.5 GHz and 5.7 GHz.
- OFDM. The 802.16 is (mainly) built on the Orthogonal Frequency Division Multiplexing (OFDM) transmission technique known for its high radio resource use efficiency.
- Data rates. A reasonable number is 10 Mb/s. Reports have given more ambitious figures going up to 70 Mb/s or even 100 Mb/s. These values would be for a very good state of the radio channel and for a very small cell capacity, making these values too optimistic for the moment.
- Distance. Up to 20 km, a little less for indoor equipments.

As mentioned in Chapter 1, the IEEE 802.16 standard is the network technology used for WiMAX. The IEEE 802.16 working group for BWA was created in 1999. It was divided into two working groups:

- 802.16a, centre frequency within the interval 2–11 GHz. This technology will then be used for WiMAX.
- 802.16, with a frequency value interval of 10–66 GHz.

Many documents were approved and published by 802.16 subcommittees. They are presented in Table 2.1.

As stated in 802.16-2004 [1], this standard revises and consolidates IEEE standards 802.16-2001, 802.16a-2003 and 802.16c-2002. Before getting to 802.16-2004, a revision called 802.16d was started in September 2003 with the objective of taking into account the ETSI HiperMAN BWA standard [3]. The 802.16d project was later concluded with the approval of the 802.16-2004 document and the withdrawal of the earlier 802.16 documents, including the a, b and c amendments. Confusingly enough, some people still refer to 802.16-2004 as 802.16d (or even 16d).

WiMAX: Technology for Broadband Wireless Access   Loutfi Nuaymi
© 2007 John Wiley & Sons, Ltd

Table 2.1 Main IEEE 802.16 documents

Date and name of the document	Description
Dec. 2001, 802.16	10–66 GHz; line-of-sight (LOS); 2–5 km; channel bandwidth values: 20, 25 and 28 MHz
Jan. 2003, 802.16a	2–11 GHz; non-line-of-sight (NLOS)
Oct. 2004, 802.16-2004	Revises and consolidates previous 802.16 standards; replaces 16a and 16; 5–50 km
7 Dec. 2005, 802.16 approves 802.16e amendment of 802.16-2004	Mobility; OFDMA (SOFDMA)
Other 802.16 amendments approved or at draft stage: 802.16f, 802.16g, 802.16f, etc.	See Section 2.5

2.1.1 From 802.16-2004 to 802.16e

802.16-2004 was definitely very useful, replacing a set of documents all describing different parts of the same technology, with different modification directions. Yet, after its publication, it still needed an upgrade, mainly for the addition of mobility features. Other features were needed and some errors had to be corrected. This gave way to 802.16e amendment approved on December 7, 2005 and published in February 2006 [2].

It should be noted that 802.16e is not a standalone document. It only proposes (sometimes important) changes and additions to the 802.16-2004 text. Hence, a person wishing to read the details of specific information in 802.16, e.g. 'What is the frame format in 802.16?' has first to read the related part of 802.16-2004 and then go on to read the possible changes that took place in 802.16e. It was reported that the IEEE intention was to have a unique document resulting from 16-2004 and 16e fusion, called 802.16-2005. However, by summer 2006, this document does not exist (to the best of the author's knowledge). However, the 802.16-2004 standard and 802.16e amendment are sometimes referred to as the IEEE 802.16-2005 standard.

The main differences of 802.16e with regard to 802.16-2004 are the following (the list is not exhaustive):

- Mobile stations (MS) appear. A station in a mobile telecommunication service is intended to be used while in motion or during halts at unspecified points. However, a 802.16e MS is also a subscriber station (SS).
- MAC layer handover procedures. There are two types of handover (see Chapter 14).
- Power save modes (for mobility-supporting MSs): sleep mode and idle mode (see Chapter 14).
- SOFDMA (Scalable OFDMA). More generally, the OFDMA PHY layer, i.e. Section 8.4 of the 802.16 standard, was completely rewritten between 16-2004 and 16e. Although the word SOFDMA does not appear in the 802.16e document, it is the type of standardised OFDMA. For OFDMA and SOFDMA, see Chapter 5.
- Security (privacy sublayer). The security of 16-2004 is completely updated (see Chapter 15).
- Multiple-Input Multiple-Output (MIMO) and Adaptive Antenna System (AAS) techniques, both already introduced in 802.16-2004, have many enhancement and implementation details provided in 802.16e (see Chapter 12).
- Multicast and broadcast services (MBS) feature.

Table 2.2   Some WiMAX Forum members

Manufacturers	Airspan, Alcatel, Alvarion, Broadcom, Cisco, Ericsson, Fujitsu, Huawei, Intel, LG, Lucent, Motorola, Navini, Nokia, Nortel, NEC, Proxim, Sagem, Samsung, Sequans, Siemens, ZTE, etc.
Service providers	British Telecom, France Telecom, KT (Korea Telecom), PCCW, Sprint Nextel, Telmex, etc.

- A new (fifth) QoS class: ertPS. (In addition to 802.16-2004 rtPS), ertPS Class supports real-time service flows that generate variable-size data packets on a periodic basis, e.g. VoIP with silence suppression.
- Other: the Low-Density Parity Check (LDPC) code is an optional channel coding, etc.

2.2 WiMAX Forum

IEEE 802 standards provide only the technology. It is then needed to have other organisms for the certification of conformity and the verification of interoperability. In the case of IEEE 802.11 WLAN, the Wireless Fidelity Alliance (WiFi or Wi-Fi) Consortium had a major role in the success of the WiFi technology, as it is now known. Indeed, the fact that two WiFi certified IEEE 802.11 WLAN devices are guaranteed to work together paved the way for the huge spread of WiFi products.

The certification problem was even more important for WiMAX as many product manu-facturers claimed they had verified the 802.16 standard (for pre-WiMAX products, see Sec-tion 1.4.2). The WiMAX (Worldwide Interoperability for Microwave Access) Forum (www. wimaxforum.org) was created in June 2001 with the objective that the WiMAX Forum plays exactly the same role for IEEE 802.16 as WiFi for 802.11. The WiMAX Forum provides certification of conformity, compatibility and interoperability of IEEE 802.16 products. After a period of low-down, the WiMAX Forum was reactivated in April 2003. Some sources indicate this latter date as the date of the creation of the WiMAX Forum. Intel and Nokia, along with others, played a leading role in the creation of the Forum. Then Nokia became less active, claiming that it wished to concentrate on 3G. However, Nokia is again an active player of WiMAX.

WiMAX Forum members are system and semiconductors manufacturers, other equipment vendors, network operators, academics and other telecommunication actors. A complete list of the WiMAX Forum members can be found on the Forum Member Roster web page. A nonexhaustive list of WiMAX members is proposed in Table 2.2.

The site of the WiMAX Forum indicates that its objective is to facilitate the deployment of broadband wireless networks based on the IEEE 802.16 standard by ensuring the compat-ibility and interoperability of broadband wireless equipment. More details about WiMAX certification are given in Section 2.3.

2.2.1 WiMAX Forum Working Groups

The WiMAX Forum is organised into Working Groups (WGs). The scope of these WGs is given in Table 2.3, as indicated on the WiMAX Forum website.

The WiMAX network architecture as defined by the NWG is described in Chapter 13.

Table 2.3 WiMAX Forum working groups. As of July 2006, the Forum website also indicates the Global Roaming Working Group (GRWG)

Working group name	Scope
Application Working Group (AWG)	Defines applications over WiMAX that are necessary to meet core competitive offerings and are uniquely enhanced by WiMAX
Certification Working Group (CWG)	Handles the operational aspects of the WiMAX Forum certification program; interfaces with the certification lab(s); selects new certification lab(s).
Marketing Working Group (MWG)	Promotes the WiMAX Forum, its brands and the standards that form the basis for worldwide interoperability of BWA systems
Network Working Group (NWG)	Creates higher-level networking specifications for fixed, nomadic, portable and mobile WiMAX systems, beyond what is defined in the scope of 802.16; specifically, the NWG defines the architecture of a WiMAX network
Regulatory Working Group (RWG)	Influences worldwide regulatory agencies to promote WiMAX-friendly, globally harmonised spectrum allocations
Service Provider Working Group (SPWG)	Gives service providers a platform for influencing BWA product and spectrum requirements to ensure that their individual market needs are fulfilled
Technical Working Group (TWG)	Develops conformance test specifications and certification services and profiles based on globally accepted practices to achieve worldwide interoperability of BWA systems

2.2.2 WiMAX Forum White Papers

The WiMAX Forum regularly publishes White Papers. These are a very useful information source about WiMAX, freely available on the Forum website. In Table 2.4, a nonexhaustive list of White Papers is proposed (until July 2006).

2.3 WiMAX Products Certification

The WiMAX forum first recognised the Centro de Tecnología de las Comunicaciones, (Cetecom Lab) (www.cetecom.es), located in Malaga, Spain, as the first certification lab of WiMAX products. In February 2006, the WiMAX Forum designated the Telecommunications Technology Association's (TTA) IT Testing and Certification Lab in Seoul, South Korea, as the second lab available to WiMAX Forum members to certify compatibility and interoperability of WiMAX products. The first certifications of this latter lab are expected in 2007. The process for selecting a third WiMAX certification lab in China has been reported.

WiMAX conformance should not be confused with interoperability [5]. The combination of these two types of testing make up certification testing. WiMAX conformance testing is a process where BS and SS manufacturers test units to ensure that they perform in accordance with the specifications called out in the WiMAX Protocol Implementation Conformance

Table 2.4 WiMAX Forum (www.wimaxforum.org) White Papers, last update: July 2006. Table was drawn with the help of Ziad Noun

Title	Date of latest version	Number of pages	Brief description
IEEE 802.16a standard and WiMAX – Igniting BWA	Date not mentioned	7	An overview of IEEE 802.16a standard, its PHY and MAC layers; talks also about the WiFi versus WiMAX scalability
Regulatory position and goals of the WiMAX Forum	August 2004	6	Describes the goals of WiMAX Forum (interoperability of broadband wireless products); describes also the initial frequency bands (license and license exempt)
Business case for fixed wireless access in emerging markets	June 2005	22	Describes the characteristics of emerging markets and discusses the service and revenue assumptions for business case analysis (urban, suburban, rural)
WiMAX deployment considerations for fixed wireless access in the 2.5 GHz and 3.5 GHz licensed bands	June 2005	21	About the licensed spectrum for WMAN, the radio characteristics, the range and the capacity of the system in different scenarios (urban, suburban, etc.)
Business case models for fixed broadband wireless access based on WiMAX technology and the 802.16 standard	October 2004	24	Describes the WiMAX architecture and applications, the business case considerations and assumptions and the services offered by WiMAX
Initial certification profiles and the European regulatory framework	September 2004	4	Describes the profiles currently identified for the initial certification process and the tentative profiles under consideration for the next round of the certification process
WiMAX's technology for LOS and NLOS environments.	August 2004	10	About the characteristics of OFDM and the other solutions used by WiMAX to solve the problems resulting from NLOS (subchannelisation, directional antennas, adaptive modulation, error correction techniques, power control, etc.)
Telephony's 'Complete Guide to WiMAX'	May 2004	10	About WiMAX marketing and policy considerations
What WiMAX Forum certified products will bring to Wi-Fi	June 2004	10	Why WiFi is used in WiMAX, the OFDM basics, the 802.16/HiperMAN PHY and MAC layers, the operator requirements for BWA systems and the products certification

(continued overleaf)

Table 2.4   (*continued*)

Title	Date of latest version	Number of pages	Brief description
What WiMAX Forum certified products will bring to 802.16	June 2004	6	The certified products: where do WiMAX Forum certified products fit and why select them?
Fixed, nomadic, portable and mobile applications for 802.16-2004 and 802.16e WiMAX networks	November 2005	16	Compares the two possibilities of deployment for an operator: fixed WiMAX (802.16-2004) or mobile WiMAX (802.16e)
The WiMAX Forum certified program for fixed WiMAX	March 2006	15	Describes the general WiMAX certification process and specifically the fixed WiMAX system profiles certifications
Third WiMAX Forum plugfest – test methodology and key learnings	March 2006	18	Describes WiMAX March 2006 plugfest
Mobile WiMAX – Part I: a technical overview and performance evaluation	March 2006	53	Technical overview of 802.16e system (mobile WiMAX) and the corresponding WiMAX architecture
Mobile WiMAX – Part II: a comparative analysis	May 2006	47	Compares elements between mobile WiMAX and presently used 3G systems (1xEVDO and HSPA)
Mobile WiMAX: the best personal broadband experience!	June 2006	19	Provides mobile WiMAX advantages in the framework of mobile broadband access market
Executive summary: mobile WiMAX performance and comparative summary	July 2006	10	Brief overview of mobile WiMAX and summary of previous White Paper performance data

Specification (PICS) documents. The WiMAX PICS documents are proposed by the TWG (see the previous section). In the conformance test, the BS/SS units must pass all mandatory and prohibited test conditions called out by the test plan for a specific system profile. The WiMAX system profiles are also proposed by the TWG.

WiMAX interoperability is a multivendor (≥ 3) test process hosted by the certification lab to test the performance of the BS and/or SS from one vendor to transmit and receive data bursts of the BS and/or SS from another vendor based on the WiMAX PICS. Then, each SS, for example, is tested with three BSs, one from the same manufacturers, the two others being from different manufacturers. A group test, formally known as a plugfest [6], is a meeting where many vendors can verify the interoperability of their equipments.

2.3.1 WiMAX Certified Products

The certification process started in the summer of 2005 in Cetecom. The first equipment certification took place on 24 January 2006. The complete list of certified WiMAX equipments can be found on www.wimaxforum.org/kshowcase/view. All these equipments were certified for IEEE 802.16-2004 profiles (fixed WiMAX). Certification of equipments based on mobile WiMAX profiles (or, soon on mobile WiMAX equipments) should take place in the first half of 2007.

The certified equipments are from the three types of WiMAX manufacturers:

- pre-WiMAX experienced companies;
- companies initially more specialised in cellular network products, e.g. Motorola, which is in these two categories;
- newcomers that started business specifically for WiMAX products.

2.4 Predicted Products and Deployment Evolution

2.4.1 Product Types

Different types of WiMAX products are expected.

First step: CPE products. These CPE products are first outdoor (see Figure 1.5) and then indoor. These are the products already certified (mainly outdoor for the moment). For CPEs WiMAX products, some providers may require that only authorised installers should install the equipment for subscribers. It can be expected that self-installed CPEs will quickly appear.

Second step: devices installed on portable equipments. These portable equipments will first be laptops. It is expected (and probably already realised by the time of publication of this book) that these laptop-installed WiMAX devices may have a USB (Universal Serial Bus) connection, PCMCIA (Personal Computer Memory Card International Association) (less probable), a PCI (Peripheral Component Interconnect) connection or another type of connection. In this case, a WiMAX subscriber can move in a limited area (the one covered by the BS) and then nomadicity will be realised.

Later, a WiMAX internal factory-installed device in laptops will probably appear, as is already the case for WiFi. This will clearly produce a situation where WiMAX will spread widely. The difficulties encountered are of two types:

- manufacturing devices small enough; this do not really seem to be a difficult problem;
- radio engineering and deployment considerations, where the technology and deployment techniques should be mature enough to have a high concentration of subscribers.

Final step: WiMAX devices in PDA and other handheld devices such as a mobile phone. For this, WiMAX devices need to be even smaller. They could take the shape of the SIM (Subscriber Identity Module) cards presently used for cellular systems (second and third generation). Thus WiMAX will be a mobile network and then a competitor for 3G systems.

2.4.2 Products and Deployment Timetable

Once WiMAX evolution is described, we need to know about the timetable of these products. What about the network deployments? As of today a large number of pre-WiMAX networks exist around the world, both in developed and developing countries. These deployments are often on a scale smaller than the whole country, typically limited to a region or an urban

Table 2.5 WiMAX products and networks timetable: (e), expected

	Products	Certification	Networks
2005	Proprietary (pre-WiMAX); outdoor CPE		Fixed
2006	Pre-WiMAX equipments; first use of WiMAX certified products	Since January 2006, certification of fixed WiMAX equipments based on IEEE 802.16-2004 (see Section 2.3.1)	Launch of WiBro service in Korea; (e) first nomadic use of WiMAX?
2007	(e) Indoor, self-installed; (e) first use of mobile WiMAX, wave 1 (no MIMO and AAS, etc.)	(e) Certification of mobile WiMAX equipments based on IEEE 802.16e	(e) Nomadic use of WiMAX
2008	(e) Ramp-up of mobile WiMAX products, wave 1 and wave 2 (MIMO and AAS)		(e) Mobility

zone. For example, in France, Altitude Telecom operator proposes a BWA subscription in four geographic departments: Calvados, Orne, Seine-et-Marne and Vendée. The displayed data rate is 1 Mb/s (June 2006). Many fixed WiMAX networks (then using the recently certified products) are imminent, some of them belonging to pre-WiMAX operators planning to upgrade to certified WiMAX.

Table 2.5 is based on documents and conferences by WiMAX actors. The (e), expected, dates are only assumptions. Some of these previewed dates may be changed in the future.

2.5 Other 802.16 Standards

In addition to the 802.16e amendment of the 802.16 standard, other amendments have been made or are still in preparation. The goal of these amendments is to improve certain aspects of the system (e.g. have a more efficient handover) or to clarify other aspects (e.g. management information).

The 802.16f amendment, entitled 'Management Information Base', was published in December 2005 and provides enhancements to IEEE 802.16-2004, defining a Management Information Base (MIB) for the MAC and PHY and the associated management procedures (see Section 3.6 for more details on 802.16f).

The 802.16g amendment was still at the draft stage in October 2006. The draft is entitled 'Management Plane Procedures and Services' and the amendment approval is planned for May 2007 (October 2006 information). It should provide the elements for efficient handover, high-performance QoS (Quality of Service) management and radio resource management procedures.

Other amendments at the draft stage are the following (from the IEEE 802.16 website, July 2006):

- 802.16/Conformance04 – Protocol Implementation Conformance Statement (PICS) proforma for frequencies below 11 GHz;
- 802.16k – Media Access Control (MAC) Bridges – Bridging of 802.16.

Amendments at the pre-draft stage are the following:

- 802.16h – Improved Coexistence Mechanisms for License-Exempt Operation;
- 802.16i – Mobile Management Information Base, where the objective is to add mobility support to the 802.16f fixed MIB standard.

Work on the 802.16j amendment draft has been reported, which concerns the Multi-hop Mobile Radio (MMR). Hence, 802.16j should provide some enhancement for the Mesh mode. The Project Authorization Request (PAR) of 802.16j was approved in March 2006.

2.6 The Korean Cousin: WiBro

South Korea has definitely an advantage in modern telecommunication networks, whether in ADSL (Asymmetric Digital Subscriber Line) or 3G figures. The TTA PG302 BWA standard was approved in June 2004 by the TTA (Telecommunications Technology Association, the Korean standardisation organisation) and is known as WiBro (Wireless Broadband). This standard has the support of leading people in the Korean telecommunication industry.

Originally sought as a competitor of WiMAX, an agreement was found by the end of 2004, while 802.16e was still under preparation, between 802.16 backers (including Intel) and WiBro backers in order to have WiBro products certified as WiMAX equipments.

WiBro licenses were assigned in Korea in January 2005. The three operators are Korea Telecom (KT), SK Telecom (SKT) and Hanaro Telecom. Pilot networks are already in place (April 2006). Relatively broad coverage public commercial offers should start before the end of 2006. WiBro planned deployments in other countries have been reported (among others, Brazil). This should give WiBro an early large-scale BWA deployment and then provide important field technical and market observations.

3

Protocol Layers and Topologies

In this chapter, the protocol layer architecture of WiMAX/802.16 is introduced. The main objectives of each sublayer are given as well as the global functions that they realise. Links are provided to the chapters of this book where each of these sublayers or procedures are described in much more detail.

3.1 The Protocol Layers of WiMAX

The IEEE 802.16 BWA network standard applies the so-called Open Systems Interconnection (OSI) network reference seven-layer model, also called the OSI seven-layer model. This model is very often used to describe the different aspects of a network technology. It starts from the Application Layer, or Layer 7, on the top and ends with the PHYsical (PHY) Layer, or Layer 1, on the bottom (see Figure 3.1).

The OSI model separates the functions of different protocols into a series of layers, each layer using only the functions of the layer below and exporting data to the layer above. For example, the IP (Internet Protocol) is in Layer 3, or the Routing Layer. Typically, only the lower layers are implemented in hardware while the higher layers are implemented in software.

The two lowest layers are then the Physical (PHY) Layer, or Layer 1, and the Data Link Layer, or Layer 2. IEEE 802 splits the OSI Data Link Layer into two sublayers named Logical Link Control (LLC) and Media Access Control (MAC). The PHY layer creates the physical connection between the two communicating entities (the peer entities), while the MAC layer is responsible for the establishment and maintenance of the connection (multiple access, scheduling, etc.).

The IEEE 802.16 standard specifies the air interface of a fixed BWA system supporting multimedia services. The Medium Access Control (MAC) Layer supports a primarily point-to-multipoint (PMP) architecture, with an optional mesh topology (see Section 3.7). The MAC Layer is structured to support many physical layers (PHY) specified in the same standard. In fact, only two of them are used in WiMAX.

The protocol layers architecture defined in WiMAX/802.16 is shown in Figure 3.2. It can be seen that the 802.16 standard defines only the two lowest layers, the PHYsical Layer and the MAC Layer, which is the main part of the Data Link Layer, with the LLC layer

WiMAX: Technology for Broadband Wireless Access   Loutfi Nuaymi
© 2007 John Wiley & Sons, Ltd

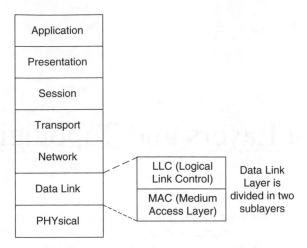

Figure 3.1   The seven-layer OSI model for networks. In WiMAX/802.16, only the two first layers are defined

very often applying the IEEE 802.2 standard. The MAC layer is itself made of three sub-layers, the CS (Convergence Sublayer), the CPS (Common Part Sublayer) and the Security Sublayer.

The dialogue between corresponding protocol layers or entities is made as follows. A Layer X addresses an XPDU (Layer X Protocol Data Unit) to a corresponding Layer X (Layer X of the peer entity). This XPDU is received as an (X-1)SDU (Layer X-1 Service Data Unit) by Layer X-1 of the considered equipment. For example, when the MAC Layer of an equipment

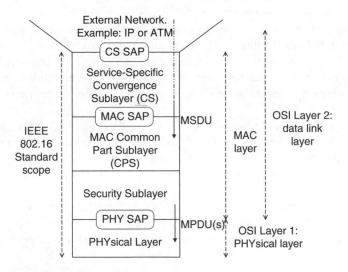

Figure 3.2   Protocol layers of the 802.16 BWA standard. (From IEEE Std. 802.16-2004 [1]. Copyright IEEE 2004, IEEE. All rights reserved.)

sends an MPDU (MAC PDU) to a corresponding equipment, this MPDU is received as a PSDU (PHYsical SDU) by the PHYsical Layer (see Figure 3.2).

In this chapter, the different layers are introduced. Each of these layers or sublayers and many of their functions are described in the following sections.

3.2 Convergence Sublayer (CS)

The service-specific Convergence Sublayer (CS), often simply known as the CS, is just above the MAC CPS sublayer (see Figure 3.2). The CS uses the services provided by the MAC CPS, via the MAC Service Access Point (SAP). The CS performs the following functions:

- Accepting higher-layer PDUs from the higher layers. In the present version of the standard [1], CS specifications for two types of higher layers are provided: the asynchronous transfer mode (ATM) CS and the packet CS. For the packet CS, the higher-layer protocols may be IP v4 (version 4) or v6 (version 6).
- Classifying and mapping the MSDUs into appropriate CIDs (Connection IDentifier). This is a basic function of the Quality of Service (QoS) management mechanism of 802.16 BWA.
- Processing (if required) the higher-layer PDUs based on the classification.
- An optional function of the CS is PHS (Payload Header Suppression), the process of suppressing repetitive parts of payload headers at the sender and restoring these headers at the receiver.
- Delivering CS PDUs to the appropriate MAC SAP and receiving CS PDUs from the peer entity.

CS procedures and operations are detailed in Chapter 7.

3.3 Medium Access Control Common Part Sublayer (MAC CPS)

The Common Part Sublayer (CPS) resides in the middle of the MAC layer. The CPS represents the core of the MAC protocol and is responsible for:

- bandwidth allocation;
- connection establishment;
- maintenance of the connection between the two sides.

The 802.16-2004 standard defines a set of management and transfer messages. The management messages are exchanged between the SS and the BS before and during the establishment of the connection. When the connection is realised, the transfer messages can be exchanged to allow the data transmission.

The CPS receives data from the various CSs, through the MAC SAP, classified to particular MAC connections. The QoS is taken into account for the transmission and scheduling of data over the PHY Layer. The CPS includes many procedures of different types: frame construction, multiple access, bandwidth demands and allocation, scheduling, radio resource management, QoS management, etc. These functions are detailed in Chapters 8 to 11.

3.4 Security Sublayer

The MAC Sublayer also contains a separate Security Sublayer (Figure 3.2) providing authentication, secure key exchange, encryption and integrity control across the BWA system. The two main topics of a data network security are data encryption and authentication. Algorithms realising these objectives should prevent all known security attacks whose objectives may be denial of service, theft of service, etc.

In the 802.16 standard, encrypting connections between the SS and the BS is made with a data encryption protocol applied for both ways. This protocol defines a set of supported cryptographic suites, i.e. pairings of data encryption and authentication algorithms. An encapsulation protocol is used for encrypting data packets across the BWA. This protocol defines a set of supported cryptographic suites, i.e. pairings of data encryption and authentication algorithms. The rules for applying those algorithms to an MAC PDU payload are also given.

An authentication protocol, the Privacy Key Management (PKM) protocol is used to provide the secure distribution of keying data from the BS to the SS. Through this secure key exchange, due to the key management protocol the SS and the BS synchronize keying data. The basic privacy mechanisms are strengthened by adding digital-certificate-based SS authentication to the key management protocol. In addition, the BS uses the PKM protocol to guarantee conditional access to network services. The 802.16e amendment defined PKMv2 which has the same framework as PKM, re-entitled PKMv1, with some additions such as new encryption algorithms, mutual authentication between the SS and the BS, support for a handover and a new integrity control algorithm.

WiMAX security procedures are described in Chapter 15.

3.5 PHYsical Layer

WiMAX is a BWA system. Hence, data are transmitted at high speed on the air interface through (radio) electromagnetic waves using a given frequency (operating frequency).

The PHY Layer establishes the physical connection between both sides, often in the two directions (uplink and downlink). As 802.16 is evidently a digital technology, the PHYsical Layer is responsible for transmission of the bit sequences. It defines the type of signal used, the kind of modulation and demodulation, the transmission power and also other physical characteristics.

The 802.16 standard considers the frequency band 2–66 GHz. This band is divided into two parts:

- The first range is between 2 and 11 GHz and is destined for NLOS transmissions. This was previously the 802.16a standard. This is the only range presently included in WiMAX.
- The second range is between 11 and 66 GHz and is destined for LOS transmissions. It is not used for WiMAX.

Five PHYsical interfaces are defined in the 802.16 standard. These physicals interfaces are summarised in Table 3.1. The five physical interfaces are each described in a specific section of the 802.16 standard (and amendments). The MAC options (AAS, ARQ, STC, HARQ, etc.) will be described further in this book (see the Index). Both major duplexing modes, Time Division Duplexing (TDD) and Frequency Division Duplexing (FDD), can be included in 802.16 systems.

For frequencies in the 10–66 GHz interval (LOS), the WirelessMAN-SC PHY is specified. For frequencies below 11 GHz (LOS) three PHYsical interfaces are proposed:

- WirelessMAN-OFDM, known as OFDM and using OFDM transmission;
- WirelessMAN-OFDMA, known as OFDMA and using OFDM transmission, and Orthogonal Frequency Division Multiple Access (OFDMA), with the OFDMA PHY Layer,

Table 3.1 The five PHYsical interfaces defined in the 802.16 standard. (From IEEE Std 802.16e-2005 [2]. Copyright IEEE 2006, IEEE. All rights reserved.)

Designation	Frequency band	Section in the standard	Duplexing	MAC options
WirelessMAN-SC (known as SC)	10–66 GHz (LOS)	8.1	TDD and FDD	
WirelessMAN-SCa (known as SCa)	Below 11 GHz (NLOS); licensed	8.2	TDD and FDD	AAS (6.3.7.6), ARQ (6.3.4), STC (8.2.1.4.3), mobility
WirelessMAN-OFDM (known as OFDM)	Below 11 GHz; licensed	8.3	TDD and FDD	AAS (6.3.7.6), ARQ (6.3.4), STC (8.3.8), mesh (6.3.6.6), mobility
WirelessMAN-OFDMA (known as OFDMA)	Below 11 GHz; licensed	8.4	TDD and FDD	AAS (6.3.7.6), ARQ (6.3.4), HARQ (6.3.17), STC (8.4.8), mobility
WirelessHUMAN	Below 11 GHz; license exempt	8.5 (in addition to 8.2, 8.3 or 8.4)	TDD only	AAS (6.3.7.6), ARQ (6.3.4), STC (see above), only with 8.3, mesh (6.3.6.6)

described in Section 8.4 of the 802.16 standard, being completely rewritten between 802.16-2004 and 802.16e;
• WirelessMAN-SCa, known as SCa and using single-carrier modulations.

Some specifications are given for the unlicensed frequency bands used for 802.16-2004 in the framework of the WirelessHUMAN (High-speed Unlicensed Metropolitan Area Network) PHYsical Layer. Unlicensed frequency is included in fixed WiMAX certification. For unlicensed frequency bands, in addition to the features mentioned in Table 3.1, the standard [2] requires mechanisms such as Dynamic Frequency Selection (DFS) to facilitate the detection and avoidance of interference and the prevention of harmful interference into other users, including specific spectrum users identified by regulations [7] as priority users.

WiMAX considers only OFDM and OFDMA PHYsical layers of 802.16 (see Figure 3.3). The PHYsical Layer is described in Chapter 7, where the OFDM transmission technique is described. Efficiency of the use of the frequency bandwidth is treated in Chapter 12.

3.5.1 Single Carrier (SC) and OFDM

The use of OFDM increases the data capacity and, consequently, the bandwidth efficiency with regard to classical Single Carrier (SC) transmission. This is done by having carriers very close to each other but still avoiding interference because of the orthogonal nature of these carriers. Therefore, OFDM presents a relatively high spectral efficiency. Numbers of the order of magnitude of 3.5–5 b/s Hz for spectral efficiency are often given. This is greater than the values often given for CDMA (Code Division Multiple Access) used for 3G, although this is not a definitive assumption as it depends greatly on the environment and other system parameters.

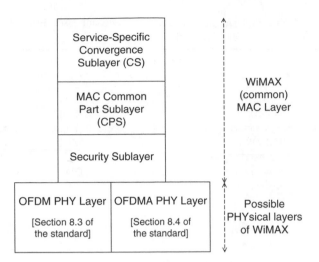

Figure 3.3   IEEE 802.16 common MAC Layer can be used with two possible PHYsical layers in WiMAX

The OFDM transmission technique and its use in OFDM and OFDMA physical layers of WiMAX are described in Chapter 5.

3.6 Network Management Reference Model

The 802.16f amendment [8] provides enhancements to IEEE 802.16-2004, defining a Management Information Base (MIB) for the MAC and PHY and the associated management procedures. This document describes the use of a Simple Network Management Protocol (SNMP), an Internet Engineering Task Force (IETF) protocol (RFCs 1902, 1903, 3411-5 and 3418), as a network management reference model.

802.16f consists of a Network Management System (NMS), managed nodes and a service flow database (see Figure 3.4). BS and SS managed nodes collect and store the managed objects in the format of WiressMan Interface MIB and wmanDevMib, defined in the 802.16f document, which are made available to NMSs via management protocols, such as the Simple Network Management Protocol (SNMP). The service flow database contains the service flow and the associated QoS information that need to be associated to the BS and the SS when an SS enters into a BS network.

3.7 WiMAX Topologies

The IEEE 802.16 standard defines two possible network topologies (see Figure 3.6):

- PMP (Point-to-Multipoint) topology (see Figure 3.5);
- Mesh topology or Mesh mode (see Figure 3.6).

The main difference between the two modes is the following: in the PMP mode, traffic may take place only between a BS and its SSs, while in the Mesh mode the traffic can be routed

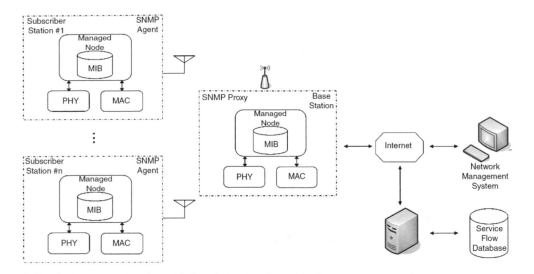

Figure 3.4 Network management reference model as defined in 802.16f. (From IEEE Std 802.16f-2005 [8]. Copyright IEEE 2005, IEEE. All rights reserved.)

through other SSs until the BS and can even take place only between SSs. PMP is a centralised topology where the BS is the centre of the system while in Mesh topology it is not. The elements of a Mesh network are called nodes, e.g. a Mesh SS is a node.

In Mesh topology, each station can create its own communication with any other station in the network and is then not restricted to communicate only with the BS. Thus, a major advantage of the Mesh mode is that the reach of a BS can be much greater, depending on the

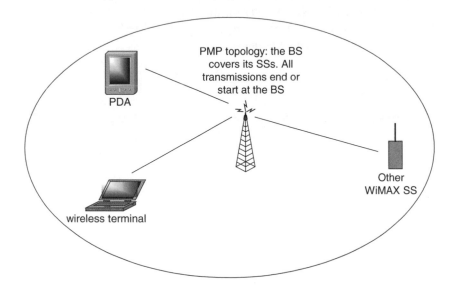

Figure 3.5 PMP topology

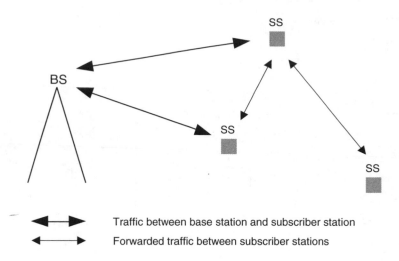

Traffic between base station and subscriber station
Forwarded traffic between subscriber stations

Figure 3.6 Mesh topology. The BS is no longer the centre of the topology, as in the classical PMP mode

number of hops, until the most distant SS. On the other hand, using the Mesh mode brings up the now thoroughly studied research topic of ad hoc (no fixed infrastructure) networks routing.

When authorised to a Mesh network, a candidate SS node receives a 16-bit Node ID (IDentifier) upon a request to an SS identified as the Mesh BS. The Node ID is the basis of node identification. The Node ID is transferred in the Mesh subheader of a generic MAC frame in both unicast and broadcast messages (see Chapter 8 for frame formats).

First WiMAX network deployments are planned to follow mainly PMP topology. Mesh topology is not yet part of a WiMAX certification profile (September 2006). It has been reported that some manufacturers are planning to include the Mesh feature in their products, even before Mesh is in a certification profile.

4

Frequency Utilisation and System Profiles

4.1 The Cellular Concept

The global objective of a wireless network is rather simple: to connect wireless users to a core network and then to the fixed network. Figure 4.1 illustrates the principle of a Public Land Mobile Network (PLMN), as defined for second-generation GSM networks.

The first wireless phone systems date back as far as the 1930s. These systems were rather basic and had very small capacity. The real boost for wireless networks came with the cellular concept invented in the Bell Labs in the 1970s. This simple but also extremely powerful concept was the following: each base station covers a cell; choose the cells small enough to reuse the frequencies (see Figure 4.2). Using this concept, it is theoretically possible to cover a geographical area as large as needed!

The cellular concept was applied to many cellular systems (mostly analogue) defined in the 1980s: AMPS (US), R2000 or Radiocom 2000 (France), TACS (UK), NMT (Scandinavian countries), etc. These systems being incompatible, a unique European cellular system was invented, GSM, which is presently used all over the world.

WiMAX applies the same principle: a BS covers the SSs of its cell. In this section, some elements of the cellular concept theory needed for WiMAX dimensioning are provided. First sectorisation is reminded.

4.1.1 Sectorisation

A base station site represents a big cost (both as investment, CAPEX, and functioning, OPEX) for a network operator. Instead of having one site per cell, which is the case for an omnidirectional BS, trisectorisation allows three BSs to be grouped in one site, thus covering three cells (see Figure 4.3). These cells are then called sectors. Three is not the only possibility. Generally, it is possible to have a sectorisation with n sectors. Yet, for practical reasons, trisectorisation is very often used. Sectorisation evidently needs directional antennas. Trisectorisation needs 120° antennas (such that the three BSs cover the 360°).

WiMAX: Technology for Broadband Wireless Access   Loutfi Nuaymi
© 2007 John Wiley & Sons, Ltd

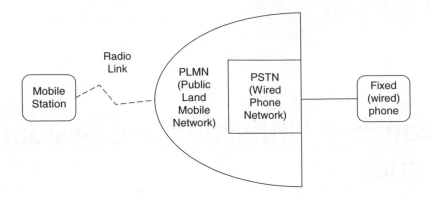

Figure 4.1   Illustration of a Public Land Mobile Network (PLMN) offering a cellular service

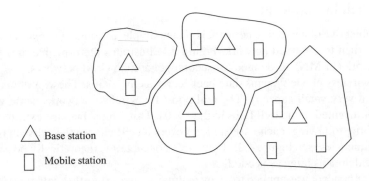

Figure 4.2   The cellular concept: simple and so powerful!

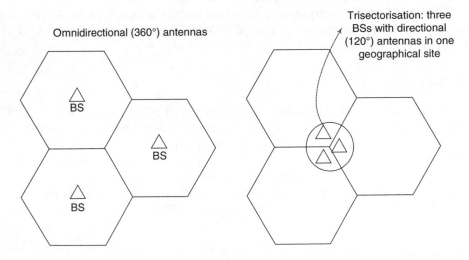

Figure 4.3   Omnidirectional antennas and trisectorisation

For economical reasons, sectorisation is almost always preferred to omnidirectional antennas for cellular networks unless in rare specific cases, e.g. for very large cells in very low populated geographic areas. WiMAX is not an exception as sectorisation is also recommended.

4.1.2 Cluster Size Considerations

Cellular networks are based on a simple principle. However, practical deployment needs a complicated planification in order to have high performance, i.e. great capacity and high quality. This planification is made with very sophisticated software tools and also the 'know-how' of radio engineers.

Frequency reuse makes room for an interference that should be kept reasonably low. As illustrated in Figure 4.4, in the downlink (for example) and in addition to its useful signal (i.e. its corresponding BS signal), an SS receives interference signals from other BSs using the same frequency. Signal-to-Noise Ratio (SNR) calculations or estimations are used for planification. The SNR is also known as the Carrier-to-Interference-and-Noise Ratio (CINR). The term SNR is used more for receiver and planification considerations, while CINR is used more for practical operations. In this book, SNR and CINR represent the same physical parameter. An appropriate cellular planification is such that the SNR remains above a fixed target value (depending on the service, among others) while maximising the capacity.

A parameter of cellular planification is the cluster size. A cluster is defined as the minimal number of cells using once and only once the frequencies of an operator (see Figure 4.5). It can be verified that having a small cluster increases the capacity per cell while big clusters decrease the global interference and then represent high quality. The choice of the cluster size must be done very carefully. In the case of GSM networks, the value of the cluster size was

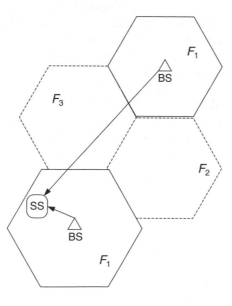

Figure 4.4 Illustration of a useful signal and interference signal as used for SNR calculations

Figure 4.5   Example of a (theoretical) regular hexagonal network. Cluster size = 3. In each cluster all the operator frequencies are used once and only once

initially 12 or 9. With time and due to different radio techniques (such as frequency hopping, among others), the cluster size of the GSM may be smaller (down to 3, possibly).

The regular hexagonal grid (as in Figure 4.5) is a model that can be used for first estimations. For this model, a relation can be established between the minimal SNR value of the network and the cluster size [9]:

$$\text{Minimal SNR} = 0.17 \times (3n)^{(\alpha/2)}$$

where n is the cluster size and α a value depending on the radio channel (of the order of 4). This is only an approximated formula used for a nonrealistic channel (too simple) and cell shapes used only for a first estimation of n. Practical deployment uses more sophisticated means but still the minimal SNR (used for cells planification or dimensioning) increases with n.

What about WiMAX frequency reuse? WiMAX is an OFDM system (while GSM is a Single Carrier, SC, system) where smaller cluster sizes can be considered. Cluster size values of 1 or 3 are regularly cited. On the other hand, it seems highly probable that sectorisation will be applied. This gives the two possible reuse schemes of Figure 4.6. Consequently, an operator having 10.5 MHz of bandwidth will have 3.5 MHz of bandwidth per cell (respectively 10.5 MHz) if a cluster of 3 (respectively 1) is considered. Yet, it is not sure that a cluster of 1 will lead to a higher global capacity. With a cluster of 1, reused frequencies are very close and then the SNR will be (globally) lower, which means less b/s Hz if link adaptation is applied (as in WiMAX, see Chapter 11). The choice of cluster size is definitely not an easy question.

Other cellular frequency reuse techniques can be also be used. Reference [10] mentions fractional frequency reuse. With an appropriate subchannel configuration, users operate on subchannels, which only occupy a small fraction of the whole channel bandwidth. The subchannel reuse pattern can be configured so that users close to the base station operate with all the subchannels (or frequencies) available, while for the edge users, each cell (or sector) operates on a

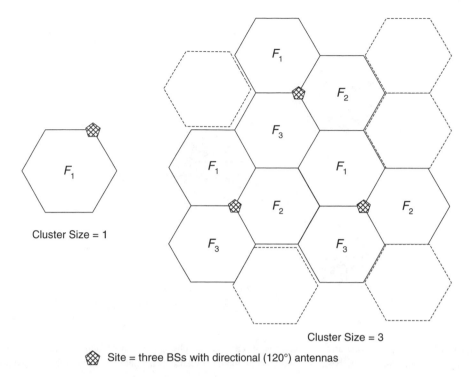

Cluster Size = 1

Cluster Size = 3

◈ Site = three BSs with directional (120°) antennas

Figure 4.6   Possible frequency reuse schemes in WiMAX

fraction of all subchannels available. With this configuration, the full-load frequency reuse is maintained for centre users in order to maximise spectral efficiency and fractional frequency reuse is implemented for edge users to assure edge user connection quality and throughput. The subchannel reuse planning can be dynamically optimised across sectors or cells based on network load and interference conditions on a frame-by-frame basis. Figure 4.7 shows an example of operating frequencies for each geographical zone in a fractional frequency reuse scheme. In the OFDMA PHYsical Layer (Mobile WiMAX), flexible subchannel reuse is facilitated by the subchannel segmentation and permutation zones. A segment is a subdivision of the available OFDMA subchannels (one segment may include all subchannels).

The tools for efficient radio resource use and other radio engineering considerations for WiMAX are described in Chapter 12.

4.1.3 Handover

Handover operation (sometimes also known as 'handoff') is the fact that a mobile user goes from one cell to another without interruption of the ongoing session (whether a phone call, data session or other). Handover is a mandatory feature of a cellular network (see Figure 4.8). Many variants exist for its implementation. Each of the known wireless systems have some differences. WiMAX handover is described in Chapter 14.

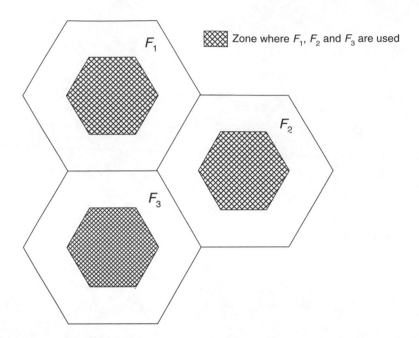

Figure 4.7 Example of operating frequencies for each geographical zone in a fractional frequency reuse scheme. The users close to the base station operate with all subchannels available. (Based on Reference [10])

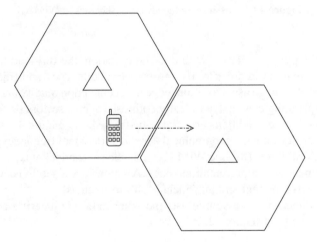

Figure 4.8   Illustration of handover in a cellular network

4.2 Licensed and Unlicensed Frequencies

3G networks have licensed fixed frequency bands. These bands are 1885–2025 GHz and 2110–2200 GHz. WiFi WLANs (11b) also have a fixed frequency band: 2410–2480 GHz. This latter band is in the ISM (Industrial, Scientific and Medical), a free band also used by systems other than WiFi.

For WiMAX, the IEEE 802.16 standard considers that the system works on frequencies smaller than 11 GHz (for the PHYsical layers considered for WiMAX). Precise operating frequencies are indicated in WiMAX Forum system profiles (see Sections 4.3 and 4.4 below).

Both types of frequency bands are part of WiMAX:

- licensed bands;
- license-exempt bands.

It is often mentioned that unlicensed frequencies of WiMAX will be used for limited coverages, campuses (enterprise or academic), particular initiatives, etc. In other words, operator revenue should come only from licensed frequencies where the service can be more easily guaranteed.

In some countries and regions, attributed licenses are known as agnostic licenses, i.e. no specific technology or other requirements are mandatory. Only frequency filter shapes and maximum transmitted power are indicated. For example, in these bands, an operator can have 3G or UMTS. However, not all attributed frequencies are agnostic. In some countries, a WiMAX licensed band may only be used for WiMAX operation. Additional constraints may also exist. For example, WiMAX cannot be used for mobile operations in some countries.

The standard indicates that the RF (Radio Frequency) centre frequency is the centre of the frequency band in which a BS or an SS may transmit. Uplink and downlink centre frequencies must be multiples of 250 kHz. The required precision is of the order of 10^{-5} (depending on the PHYsical Layer and whether it is the BS or the SS).

4.2.1 Frequency Channels and Spectral Masks

The spectral mask of the interfering signal of WiMAX working on licensed bands depends on local regulatory requirements (in Europe, ETSI requirements). When changing from one frequency (burst profile) to another, margins must be maintained to prevent saturation of the amplifier and to prevent violation of emission masks.

Concerning license-exempt bands ([1], Section 8.5), the channel centre frequency is given by

$$\text{Channel centre frequency (MHz)} = 5000 + 5n_{ch}$$

where $n_{ch} = 0,1,...,199$ is the channel number. This definition provides a numbering system for all channels, with 5 MHz spacing, from 5 GHz to 6 GHz. The standard indicates that this provides flexibility to define channelisation sets for current and future regulatory domains.

Channelisation has been found to be compatible with the WiFi WLAN 802.11a variant, for interference mitigation purposes, in the 5 GHz (US definition) Unlicensed National Information Infrastructure (U-NII) frequency band (see Table 4.1).

In license-exempt bands, the transmitted spectral density of the transmitted signal must fall within the spectral mask of Figure 4.9. The measurements must be made using the 100 kHz resolution bandwidth. The 0 dBr level is the maximum power allowed by the relevant regulatory body.

Table 4.1   License-exempt band channels. Current applicable regulations do not allow this standard to be operated in the CEPT band B. (From IEEE Std 802.16-2004 [1]. Copyright IEEE 2004, IEEE. All rights reserved.)

Regulatory domain	Band (GHz)	Channel Number	
		20 MHz channels	10 MHz channels
USA	U-NII middle 5.25–5.35	56, 60, 64	55, 57, 59, 61, 63, 65, 67
	U-NII upper 5.725–5.825	149, 153, 157, 161, 165	148, 150, 152, 154, 156, 158, 160, 162, 164, 166
Europe	CEPT band B 5.47–5.725	100, 104, 108, 112, 116, 120, 124, 128, 132, 136	99, 101, 103, 105, 107, 109, 111, 113, 115, 117, 119, 121, 123, 125, 127, 129, 131, 133, 135, 137
	CEPT band C 5.725–5.875	148, 152, 156, 160, 164, 168	147, 149, 151, 153, 155, 157, 159, 161, 163, 165, 167, 169

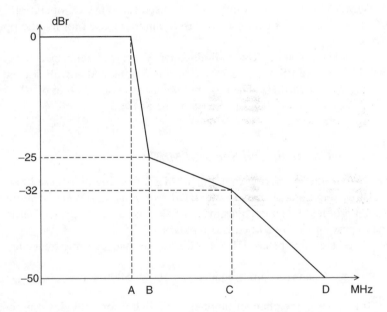

Figure 4.9   Transmit spectral mask (see also Table 4.2). (From IEEE Std 802.16-2004 [1]. Copyright IEEE 2004, IEEE. All rights reserved.)

4.3 WiMAX Frequencies, Regulations and Availability

In this section, some of the frequencies that are expected to be used for WiMAX are given. The frequency bands that will be used in one country or another for the moment (October 2006) are:

- Licensed bands: 2.3 GHz, 2.5 GHz (remember that the 2.4 GHz band is a free band used, among others, by WiFi), 3.3 GHz and 3.5 GHz, the latter being the most (geographically)

Table 4.2 Transmit spectral mask parameters [1]. A, B, C and D are in MHz

Channelisation (MHz)	A	B	C	D
20	9.5	10.9	19.5	29.5
10	4.75	5.45	9.75	14.75

widely announced WiMAX frequency band. We here mention that third-generation (3G) cellular systems operating in the 2.5 GHz band as an extension band for these systems have been reported.

- License-exempt bands: 5 GHz. The 2004 WiMAX unlicensed frequency fixed profile used the upper U-NII frequency band, i.e. the 5.8 GHz frequency band (see Table 4.1). In the future, various bands between 5 GHz and 6 GHz can be used for unlicensed WiMAX, depending on the country involved.

Table 4.3 shows (globally) the present expected WiMAX frequencies around the world. Other frequencies are sought. These frequencies should not be higher than the 5.8 GHz already chosen because, for relatively high frequencies (3.5 GHz is itself not a very small value), NLOS operation becomes difficult, which is an evident problem for mobility. The Regulatory Working Group (RWG), introduced in Chapter 2, is trying to define both new frequencies (reports talk about 450 MHz and 700 MHz) and also the conditions for an easy universal roaming with (possible) different frequencies in different countries. Regulator requirements mainly allow both Time Division Duplexing (TDD) and Frequency Division Duplexing (FDD). The attributed frequency spectrum size is a function of the country. Some elements about the WiMAX situation in some countries are given below.

4.3.1 France

In France, as elsewhere, the authorities wish to have (at least fixed) broadband access in the highest possible percentage of the territory. WiMAX has been seen as a means to provide this broadband access. Altitude Operator (owned by Iliad) has a WiMAX license in the 3.5 GHz band. Altitude obtained it in 2003 when the regulating authority, Autorité de Régulation des Télécommunication (ART), accepted that Altitude takes a WLL license owned (and not used) by another operator. Since then, ART has changed its name to become ARCEP (Autorité de Régulation des Communications Electroniques et des Postes, http://www.arcep.fr).

Table 4.3 Expected WiMAX frequencies (based on RWG documents)

Region or country	Reported WiMAX frequency bands
USA	2.3, 2.5 and 5.8 GHz
Central and South America	2.5, 3.5 and 5.8 GHz
Europe	3.5 and 5.8 GHz; possible: 2.5 GHz
South-East Asia	2.3, 2.5, 3.3, 3.5 and 5.8 GHz
Middle East and Africa	3.5 and 5.8 GHz

In August 2005, ARCEP started the process of attribution of two other WiMAX licenses (2×15 MHz each):

- BLR 1: 3465–3480 and 3565–3580 MHz;
- BLR 2: 3432.5–3447.5 and 3532.5–3547.5 MHz.

This process ended in July 2006 by the allocation of these two licences to two operators in each of the 22 French metropolitan regions. However, Altitude is the only French operator with a national WiMAX license. The choice was made based on three equally important criteria:

- contribution to the territorial development of broadband access;
- aptitude to ameliorate a high data rate concurrence;
- allowances paid by the operator.

The operators should have a minimum number (in total) of 3500 WiMAX sites by June 2008. They will be paying 125 million euros in 2006.

4.3.2 Korea

In Korea, the frequencies attributed to WiBro are in the 2.3–2.4 GHz band. In 2002, 100 MHz bands were decided for WiBro in Korea and WiBro licenses were attributed in January 2005. The three operators are Korea Telecom (KT), SK Telecom (SKT) and Hanaro Telecom. Pilot networks are already in place (April 2006). Relatively broad coverage public commercial offers should start before the end of 2006.

4.3.3 USA

In the USA, a large number of 2.5 GHz band licenses (the BRS, or Broadband Radio Service, and the EBS, or Educational Broadband Service) and 2.3 GHz band licenses (WCS, or Wireless Communications Service) are owned by many operators. Sprint and Nextel have joined forces, providing them with by far the greatest number of population served by their license. In the USA, until now the 2.5 GHz band had often been attributed for the MMDS. However, EBS licenses have been given to educational entities so that they can be used for educational purposes and the Federal Communications Commission (FCC) has allowed EBS license holders to lease spectra to commercial entities under certain conditions.

4.3.4 UK

Currently, two operators have BWA licenses in the UK: PCCW (UK Broadband) and Pipex. Their licenses are in the 3.4 GHz (PCCW) and 3.5 GHz (Pipex) bands. A number of smaller operators use or plan to use a license-exempt WiMAX frequency band for limited operations.

4.3.5 China

China is a country with big dimensions and a still developing telecommunications network. For the moment (October 2006), no license for commercial service of WiMAX has been allocated. However, WiMAX trials are taking place in many regions and are regularly reported. Leading Chinese telecommunications equipment suppliers, Huawei and ZTE, are reported to be active in the WiMAX field (members of the WiMAX Forum, contributing to experiments, preparing WiMAX products, etc.).

4.3.6 Brazil

Brazil is another country with high expectations for WiMAX. Auction of 3.5 GHz and 10 GHz BWA spectra were launched in July 2006. Expectations about the possible use of the 2.5 GHz band for WiMAX have been reported.

4.4 WiMAX System Profiles

A WiMAX system certification profile is a set of features of the 802.16 standard, selected by the WiMAX Forum, that is required or mandatory for these specific profiles. This list sets, for each of the certification profiles of a system profiles release, the features to be used in typical implementation cases. System certification profiles are defined by the TWG in the WiMAX Forum. The 802.16 standard indicates that a system (certification) profile consists of five components: MAC profile, PHY profile, RF profile, duplexing selection (TDD or FDD) and power class. The frequency bands and channel bandwidths are chosen such that they cover as much as possible of the worldwide spectra allocations expected for WiMAX.

Equipments can then be certified by the WiMAX Forum according to a specific system certification profile. Two types of system profiles are defined: fixed and mobile. These profiles are introduced in the following sections.

4.4.1 Fixed WiMAX System Profiles

Table 4.4 shows the fixed WiMAX profiles [11]. These system profiles are based on the OFDM PHYsical Layer IEEE 802.16-2004 (in fact, this PHY Layer did not change very much with 802.16e). All of the profiles use the PMP mode. This was the first set of choices decided in June 2004 (at the same time as approval of IEEE 802.16-2004). Each certification profile has an identifier for use in documents such as PICS proforma statements. Further system profiles should be defined reflecting regulatory (band opportunities) and market development. Among others, new fixed certification profiles should be approved before the end of 2006. It is planned that WiMAX system profiles with a 5 MHz channel bandwidth

Table 4.4 Fixed WiMAX certification profiles, all using the OFDM PHY and the PMP modes

Frequency band (GHz)	Duplexing mode	Channel bandwidth (MHz)	Profile name
3.5	TDD	7	3.5T1
3.5	TDD	3.5	3.5T2
3.5	FDD	3.5	3.5F1
3.5	FDD	7	3.5F2
3.5	TDD	10	5.8T

Table 4.5 Release 1 Mobile WiMAX certification profiles, all using the OFDMA PHY and the PMP modes

Frequency band (GHz)	Duplexing mode	Channel bandwidth and FFT size (number of OFDMA subcarriers)
2.3–2.4	TDD	5 MHz, 512; 8.75 MHz, 1024; 10 MHz, 1024
2.305–2.320	TDD	3.5 MHz, 512; 5 MHz, 512; 10 MHz, 1024
2.496–2.690	TDD	5 MHz, 512; 10 MHz, 1024
3.3–3.4	TDD	5 MHz, 512; 7 MHz, 1024; 10 MHz, 1024
3.4–3.8	TDD	5 MHz, 512; 7 MHz, 1024; 10 MHz, 1024

and 2.5 GHz frequency band schemes will be added. Fixed certification profiles, based on 802.16e, are also planned.

4.4.2 Mobile WiMAX System Profiles

Along with the work on the 802.16e amendment, the mobile WiMAX system profiles were defined. These certification profiles, known as Release-1 Mobile WiMAX system profiles and shown in Table 4.5, were approved in February 2006. They are based on the OFDMA PHYsical Layer (IEEE 802.16e) and all include only the PMP topology. These profiles are defined by the Mobile Task Group (MTG), a subgroup of the TWG in the WiMAX Forum. Release 1 certification will probably be separated in different Certification Waves, starting with Wave 1 having only part of all Release 1 features.

In the OFDMA PHYsical Layer as amended in 802.16e, the number of OFDMA subcarriers (equivalent to the FFT size, see the next chapter) is scalable. OFDMA of WiMAX is called scalable OFDMA. The TDD mode is the only one that has been chosen for this first set, one of the reasons being that it is more resource-use efficient. FDD profiles may be defined in the future. The frame length is equal to 5 ms. Other technical aspects of the selected profiles will be introduced in the following chapters.

Part Two

WiMAX Physical Layer

5

Digital Modulation, OFDM and OFDMA

5.1 Digital Modulations

As for all recent communication systems, WiMAX/802.16 uses digital modulation. The now well-known principle of a digital modulation is to modulate an analogue signal with a digital sequence in order to transport this digital sequence over a given medium: fibre, radio link, etc. (see Figure 5.1). This has great advantages with regard to classical analogue modulation: better resistance to noise, use of high-performance digital communication and coding algorithms, etc.

Many digital modulations can be used in a telecommunication system. The variants are obtained by adjusting the physical characteristics of a sinusoidal carrier, either the frequency, phase or amplitude, or a combination of some of these. Four modulations are supported by the IEEE 802.16 standard: BPSK, QPSK, 16-QAM and 64-QAM. In this section the modulations used in the OFDM and OFDMA PHYsical layers are introduced with a short explanation for each of these modulations.

5.1.1 Binary Phase Shift Keying (BPSK)

The BPSK is a binary digital modulation; i.e. one modulation symbol is one bit. This gives high immunity against noise and interference and a very robust modulation. A digital phase modulation, which is the case for BPSK modulation, uses phase variation to encode bits: each modulation symbol is equivalent to one phase. The phase of the BPSK modulated signal is π or $-\pi$ according to the value of the data bit. An often used illustration for digital modulation is the constellation. Figure 5.2 shows the BPSK constellation; the values that the signal phase can take are 0 or π.

5.1.2 Quadrature Phase Shift Keying (QPSK)

When a higher spectral efficiency modulation is needed, i.e. more b/s/Hz, greater modulation symbols can be used. For example, QPSK considers two-bit modulation symbols.

WiMAX: Technology for Broadband Wireless Access   Loutfi Nuaymi
© 2007 John Wiley & Sons, Ltd

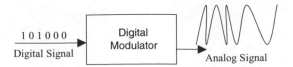

Figure 5.1 Digital modulation principle

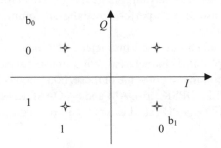

Figure 5.2 The BPSK constellation

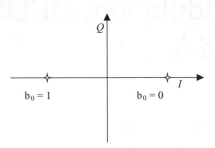

Figure 5.3 Example of a QPSK constellation

Table 5.1 shows the possible phase values as a function of the modulation symbol. Many variants of QPSK can be used but QPSK always has a four-point constellation (see Figure 5.3). The decision at the receiver, e.g. between symbol '00' and symbol '01', is less easy than a decision between '0' and '1'. The QPSK modulation is therefore less noise-resistant than BPSK as it has a smaller immunity against interference. A well-known

Table 5.1 Possible phase values for QPSK modulation

Even bits	Odd bits	Modulation symbol	φ_k
0	0	00	$\pi/4$
1	0	01	$3\pi/4$
1	1	11	$5\pi/4$
0	1	10	$7\pi/4$

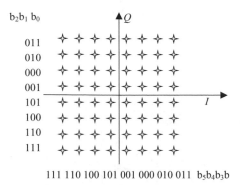

Figure 5.4   A 64-QAM constellation

digital communication principle must be kept in mind: 'A greater data symbol modulation is more spectrum efficient but also less robust.'

5.1.3 Quadrature Amplitude Modulation (QAM): 16-QAM and 64-QAM

The QAM changes the amplitudes of two sinusoidal carriers depending on the digital sequence that must be transmitted; the two carriers being out of phase of $+\pi/2$, this amplitude modulation is called quadrature. It should be mentioned that according to digital communication theory, QAM-4 and QPSK are the same modulation (considering complex data symbols). Both 16-QAM (4 bits/modulation symbol) and 64-QAM (6 bits/modulation symbol) modulations are included in the IEEE 802.16 standard. The 64-QAM is the most efficient modulation of 802.16 (see Figure 5.4). Indeed, 6 bits are transmitted with each modulation symbol.

The 64-QAM modulation is optional in some cases:

- license-exempt bands, when the OFDM PHYsical Layer is used
- for OFDMA PHY, yet the Mobile WiMAX profiles indicates that 64-QAM is mandatory in the downlink.

5.1.4 Link Adaptation

Having more than one modulation has a great advantage: link adaptation can be used (this process is also used in almost all other recent communication systems such as GSM/EDGE, UMTS, WiFi, etc.). The principle is rather simple: when the radio link is good, use a high-level modulation; when the radio link is bad, use a low-level, but also robust, modulation. Figure 5.5 shows this principle, illustrating the fact that the radio channel is better when an SS is close to the BS. Another dimension is added to this figure when the coding rate is also changed (see below).

5.2 OFDM Transmission

In 1966, Bell Labs proposed the Orthogonal Frequency Division Multiplexing (OFDM) patent. Later, in 1985, Cimini suggested its use in mobile communications. In 1997, ETSI included OFDM in the DVB-T system. In 1999, the WiFi WLAN variant IEEE 802.11g

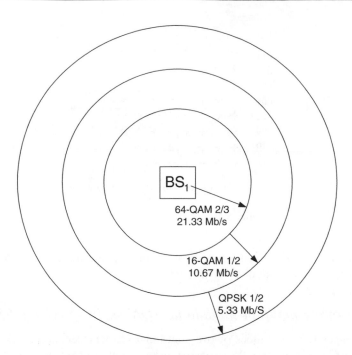

Figure 5.5 Illustration of link adaptation. A good radio channel corresponds to a high-efficiency Modulation and Coding Scheme (MCS)

considered OFDM for its PHYsical Layer. The purpose of this chapter is not to provide a complete reference for the OFDM theory and the associated mathematical proofs. Rather, the aim is to introduce the basic results needed for a minimum understanding of WiMAX.

OFDM is a very powerful transmission technique. It is based on the principle of transmitting simultaneously many narrow-band orthogonal frequencies, often also called OFDM subcarriers or subcarriers. The number of subcarriers is often noted N. These frequencies are orthogonal to each other which (in theory) eliminates the interference between channels. Each frequency channel is modulated with a possibly different digital modulation (usually the same in the first simple versions). The frequency bandwidth associated with each of these channels is then much smaller than if the total bandwidth was occupied by a single modulation. This is known as the Single Carrier (SC) (see Figure 5.6). A data symbol time is N times longer, with OFDM providing a much better multipath resistance.

Having a smaller frequency bandwidth for each channel is equivalent to greater time periods and then better resistance to multipath propagation (with regard to the SC). Better resistance to multipath and the fact that the carriers are orthogonal allows a high spectral efficiency. OFDM is often presented as the best performing transmission technique used for wireless systems.

5.2.1 Basic Principle: Use the IFFT Operator

The FFT is the Fast Fourier Transform operator. This is a matrix computation that allows the discrete Fourier transform to be computed (while respecting certain conditions). The

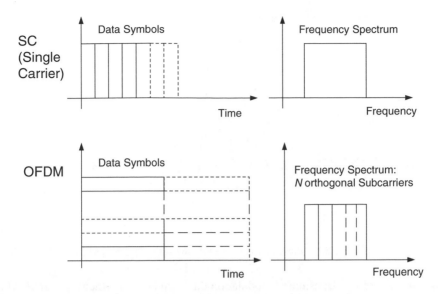

Figure 5.6 Time and frequency representation of the SC and OFDM. In OFDM, N data symbols are transmitted simultaneously on N orthogonal subcarriers

FFT works for any number of points. The operation is simpler when applied for a number N which is a power of 2 (e.g. $N = 256$). The IFFT is the Inverse Fast Fourier Transform operator and realises the reverse operation. OFDM theory (see, for example, Reference [12]) shows that the IFFT of magnitude N, applied on N symbols, realises an OFDM signal, where each symbol is transmitted on one of the N orthogonal frequencies. The symbols are the data symbols of the type BPSK, QPSK, QAM-16 and QAM-64 introduced in the previous section. Figure 5.7 shows an illustration of the simplified principle of the generation of an OFDM signal. In fact, generation of this signal includes more details that are not shown here for the sake of simplicity.

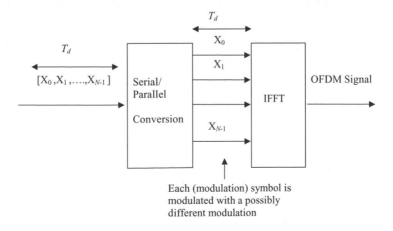

Figure 5.7 Generation of an OFDM signal (simplified)

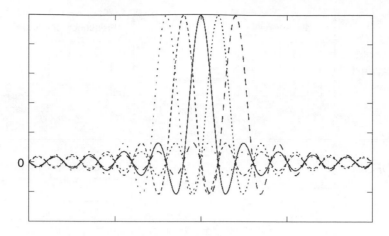

Figure 5.8   Presentation of the OFDM subcarrier frequency

If the duration of one transmitted modulation data symbol is T_d, then $T_d = 1/\Delta f$, where Δf is the frequency bandwidth of the orthogonal frequencies. As the modulation symbols are transmitted simultaneously,

$$T_d = \text{duration of one OFDM symbol}$$
$$= \text{duration of one transmitted modulation data symbol.}$$

This duration, Δf, the frequency distance between the maximums of two adjacent OFDM subcarriers, can be seen in Figure 5.8. This figure shows how the neighbouring OFDM subcarriers have values equal to zero at a given OFDM subcarrier maximum, which is why they are considered to be orthogonal. In fact, duration of the real OFDM symbol is a little greater due to the addition of the Cyclic Prefix (CP).

5.2.2 Time Domain OFDM Considerations

After application of the IFFT, the OFDM theory requires that a Cyclic Prefix (CP) must be added at the beginning of the OFDM symbol (see Figure 5.9). Without getting into

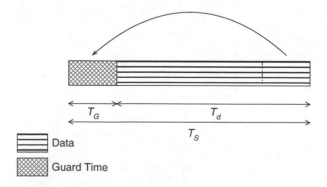

Figure 5.9   Cyclic Prefix insertion in an OFDM symbol

mathematical details of OFDM, it can be said that the CP allows the receiver to absorb much more efficiently the delay spread due to the multipath and to maintain frequency orthogonality. The CP that occupies a duration called the Guard Time (GT), often denoted T_G, is a temporal redundancy that must be taken into account in data rate computations. The ratio T_G/T_d is very often denoted G in WiMAX/802.16 documents. The choice of G is made according to the following considerations: if the multipath effect is important (a bad radio channel), a high value of G is needed, which increases the redundancy and then decreases the useful data rate; if the multipath effect is lighter (a good radio channel), a relatively smaller value of G can be used. For OFDM and OFDMA PHY layers, 802.16 defined the following values for G: 1/4, 1/8, 1/16 and 1/32. For the mobile (OFDMA) WiMAX profiles presently defined, only the value 1/8 is mandatory. The standard indicates that, for OFDM and OFDMA PHY layers, an SS searches, on initialization, for all possible values of the CP until it finds the CP being used by the BS. The SS then uses the same CP on the uplink. Once a specific CP duration has been selected by the BS for operation on the downlink, it cannot be changed. Changing the CP would force all the SSs to resynchronize to the BS [1].

5.2.3 Frequency Domain OFDM Considerations

All the subcarriers of an OFDM symbol do not carry useful data. There are four subcarrier types (see Figure 5.10):

- Data subcarriers: useful data transmission.
- Pilot subcarriers: mainly for channel estimation and synchronisation. For OFDM PHY, there are eight pilot subcarriers.
- Null subcarriers: no transmission. These are frequency guard bands.
- Another null subcarrier is the DC (Direct Current) subcarrier. In OFDM and OFDMA PHY layers, the DC subcarrier is the subcarrier whose frequency is equal to the RF centre frequency of the transmitting station. It corresponds to frequency zero (Direct Current) if the FFT signal is not modulated. In order to simplify Digital-to-Analogue and Analogue-to-Digital Converter operations, the DC subcarrier is null.

In addition, subcarriers used for PAPR reduction (see below), if present, are not used for data transmission.

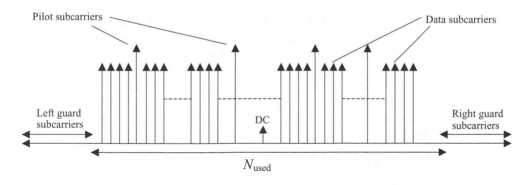

Figure 5.10   WiMAX OFDM subcarriers types. (Based on Reference [10].)

5.2.4 OFDM Symbol Parameters and Some Simple Computations

The main WiMAX OFDM symbol parameters are the following:

- The total number of subcarriers or, equivalently, the IFFT magnitude. For OFDM PHY, $N_{FFT} = 256$, the number of lower-frequency guard subcarriers is 28 and the number of higher-frequency guard subcarriers is 27. Considering also the DC subcarrier, there remains N_{used}, the number of used subcarriers, excluding the null subcarriers. Hence, $N_{used} = 200$ for OFDM PHY, of which 192 are used for useful data transmission, after deducing the pilot subcarriers.
- BW, the nominal channel bandwidth
- n, the sampling factor.

The sampling frequency, denoted f_s, is related to the occupied channel bandwidth by the following (simplified) formula:

$$f_s = n \, BW.$$

This is a simplified formula because, according to the standard, f_s is truncated to an 8 kHz multiple. According to the 802.16 standard, the numerical value of n depends of the channel bandwidths. Possible values are 8/7, 86/75, 144/125, 316/275 and 57/50 for OFDM PHY and 8/7 and 28/25 for OFDMA PHY.

5.2.4.1 Duration of an OFDM Symbol

Based on the above-defined parameters, the time duration of an OFDM symbol can be computed:

$$
\begin{aligned}
\text{OFDM symbol duration} &= \text{useful symbol time} + \text{guard time (CP time)} \\
&= 1/(\text{one subcarrier spacing}) + G \times \text{useful symbol time} \\
&= (1/\Delta f)\,(1+G) \\
&= [1/(f_s / N_{FFT})]\,(1+G) \\
&= [1/(n\,BW / N_{FFT})]\,(1+G).
\end{aligned}
$$

The OFDM symbol duration is a basic parameter for data rate computations (see below).

5.2.4.2 Data Rate Values

In OFDM PHY, one OFDM symbol represents 192 subcarriers, each transmitting a modulation data symbol (see above). One can then compute the number of data transmitted for the duration of an OFDM symbol (which value is already known). Knowing the coding rate, the number of uncoded bits can be computed. Table 5.2 shows the data rates for different Modulation and Coding Schemes (MCSs) and G values. The occupied bandwidth considered is 7 MHz and the sampling factor is 8/7 (the value corresponding to 7 MHz according to the standard).

Consider the following case in Table 5.2: 16-QAM, coding rate = 3/4 and $G = 1/16$. It can be verified that the data rate is equal to:

$$
\begin{aligned}
\text{Data rate} &= \text{number of uncoded bits per OFDM symbol/OFDM symbol duration} \\
&= 192 \times 4 \times (3/4)/\{[256/(7\,\text{MHz} \times 8/7)]\,(1 + 1/16)\} \\
&= 16.94\,\text{Mb/s}.
\end{aligned}
$$

Table 5.2 OFDM PHY data rates in Mb/s. (From IEEE Std 802.16-2004 [1]. Copyright IEEE 2004, IEEE. All rights reserved.)

G ratio	BPSK 1/2	QPSK 1/2	QPSK 3/4	16-QAM 1/2	16-QAM 3/4	64-QAM 2/3	64-QAM 3/4
1/32	2.92	5.82	8.73	11.64	17.45	23.27	26.18
1/16	2.82	5.65	8.47	11.29	16.94	22.59	25.41
1/8	2.67	5.33	8.00	10.67	16.00	21.33	24.00
1/4	2.40	4.80	7.20	9.60	14.40	19.20	21.60

It should be noted here that these data rate values do not take into account some overheads such as preambles (of the order of one or two OFDM symbols per frame) and signalling messages present in every frame (see Chapter 9 and others in this book). Hence these data rates, known as raw data rates, are optimistic values.

5.2.5 Physical Slot (PS)

The Physical Slot (PS) is a basic unit of time in the 802.16 standard. The PS corresponds to four (modulation) symbols used on the transmission channel. For OFDM and OFDMA PHY Layers, a PS (duration) is defined as [1]

$$PS = 4/f_s.$$

Therefore the PS duration is related to the system symbol rate.

This unit of time defined in the standard allows integers to be used while referring to an amount of time, e.g. the definition of transition gaps (RTG and TTG) between uplink and downlink frames in the TDD mode.

5.2.6 Peak-to-Average Power Ratio (PAPR)

A disadvantage of an OFDM transmission is that it can have a high Peak-to-Average Power Ratio (PAPR), relative to a single carrier transmission. The PAPR is the peak value of transmitted subcarriers to the average transmitted signal. A high PAPR represents a hard constraint for some devices (such as amplifiers). Several solutions are proposed for OFDM PAPR reduction, often including the use of some subcarriers for that purpose. These subcarriers are then no longer used for data transmission. The 802.16 MAC provides the means to reduce the PAPR. PAPR reduction sequences are proposed in Reference [2].

5.3 OFDMA and Its Variant SOFDMA

5.3.1 Using the OFDM Principle for Multiple Access

The OFDM transmission mode was originally designed for a single signal transmission. Thus, in order to have multiple user transmissions, a multiple access scheme such as TDMA or FDMA has to be associated with OFDM. In fact, an OFDM signal can be made from many user signals, giving the OFDMA (Orthogonal Frequency Division Multiple Access) multiple access.

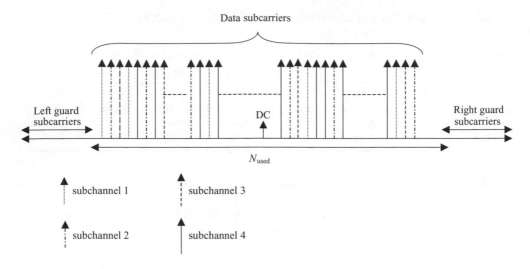

Figure 5.11   Illustration of the OFDMA principle. (Based on Reference [1].)

In OFDMA, the OFDMA subcarriers are divided into subsets of subcarriers, each subset representing a subchannel (see Figure 5.11). In the downlink, a subchannel may be intended for different receivers or groups of receivers; in the uplink, a transmitter may be assigned one or more subchannels. The subcarriers forming one subchannel may be adjacent or not. The standard [1] indicates that the OFDM symbol is divided into logical subchannels to support scalability, multiple access and advanced antenna array processing capabilities. The multiple access has a new dimension with OFDMA. A downlink or an uplink user will have a time and a subchannel allocation for each of its communications (see Figure 5.12). Different subchannel

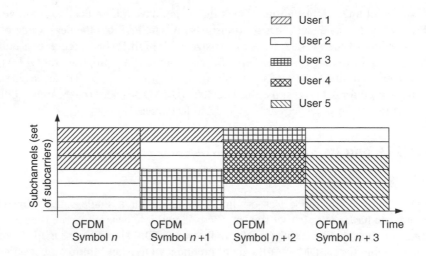

Figure 5.12   Illustration of OFDMA multiple access

Table 5.3 Example of SOFDMA figures. (Inspired from Reference [10].)

Parameters	Numerical values	
Subcarrier frequency spacing	10.95 kHz	
Useful symbol duration ($T_d = 1/\Delta f$)	91.4 µs	
Guard time ($T_G = T_d/8$)	11.4 µs	
OFDMA symbol duration ($T_s = T_d + T_G$)	102.9 µs	
Number of OFDMA symbols in the 5 ms frame	48	
FFT size (N_{FFT}) or number of subcarriers	512	1024
Channel occupied bandwidth	5 MHz	10 MHz

distributions and logical renumberings are defined in the 802.16 standard, as will be seen in the rest of this chapter. First, the SOFDMA concept is introduced.

5.3.2 Scalable OFDMA (SOFDMA)

OFDMA multiple access is not the only specificity of OFDMA PHY. Another major difference is the fact that its OFDM transmission is scalable. Although this word does not appear in the standard, OFDMA PHY is said to have Scalable OFDMA (SOFDMA). The scalability is the change of the FFT size and then the number of subcarriers. The supported FFT sizes are 2048, 1024, 512 and 128. FFT size 256 (of the OFDM layer) is not included in the OFDMA layer. Only 1024 and 512 are mandatory for mobile WiMAX profiles.

The change in the number of subcarriers, for a fixed subcarrier spacing, provides for an adaptive occupied frequency bandwidth and, equivalently, an adaptive data rate, as shown in the following example. See the example shown in Table 5.3. In this example, the sampling factor is equal to 28/25, chosen according to the channel bandwidth. SOFDMA provides an additional resource allocation flexibility that can be used in the framework of radio resource management policy taking into account the dynamic spectrum demand, among others.

5.3.3 OFDMA in the OFDM PHYsical Layer: Subchannelisation

As a matter of fact, the OFDM PHY includes some OFDMA access. Subchannelisation was included in 802.16-2004 for the uplink and also for the downlink in amendment 802.16e. The principle is the following. The 192 useful data OFDM subcarriers of OFDM PHY are distributed in 16 subchannels made of 12 subcarriers each. Each subchannel is made of four groups of three adjacent subchannels each (see below).

A subchannelised transmission is a transmission on only part of the OFDM subcarrier space. The subchannelised transmission can take place on 1, 2, 4, 8 or 16 subchannels. A five-bit indexation shown in Table 5.4 indicates the number of subchannels and the subcarrier indices used for each subchannel index for the uplink. As shown in this table, one or more pilot subcarrier(s) (there are eight in total) are allocated only if two or more subchannels are allocated. The subcarriers other than the ones used for subchannelised transmission are nonactive (for the transmitter). The five-bit subchannel index is used in the uplink allocation message UL-MAP (see Chapter 9 for the UL-MAP).

Table 5.4   The number of subchannels and the subcarrier indices used for each (five bits) subchannel index. (Based on Reference [1].)

Subchannel index				Pilot frequency index	Subchannel index (continued)	Subcarrier frequency indices
0b10000 (no subchannelisation	0b01000	0b00100	0b00010	−38	0b00001	−100:−98; −37:−35; 1:3; 64:66
					0b00011	−97:−95, −34:−32, 4:6, 67:69
			0b00110	13	0b00101	−94:−92, −31:−29, 7:9, 70:72
					0b00111	−91:−89, −28:−26, 10:12, 73:75
		0b01100	0b01010	−88	0b01001	−87:−85, −50:−48, 14:16, 51:53
					0b01011	−84,−82, −47:−45, 17:19, 54:56
			0b01110	63	0b01101	−81:−79, −44:−42, 20:22, 57:59
					0b01111	−78:−76, −41:−39, 23:25, 60:62
	0b11000	0b10100	0b10010	−13	0b10001	−75:−73, −12:−10, 26:28, 89:91
					0b10011	−72:−70, −9: −7, 29:31, 92:94
			0b10110	38	0b10101	−69:−67, −6: −4, 32:34, 95:97
					0b10111	−66:−64, −3: −1, 35:37, 98:100
		0b11100	0b11010	−63	0b11001	−62:−60, −25:−23, 39:41, 76:78
					0b11011	−59:−57, −22:−20, 42:44, 79:81
			0b11110	88	0b11101	−56:−54, −19:−17, 45:47, 82:84
					0b11111	−53:−51, −16:−14, 48:50, 85:87

Subchannelised transmission in the uplink is an option for an SS. It can be used only if the BS signals its capability to decode such transmissions. The BS must not assign to any given SS two or more overlapping subchannelised allocations in the same time.

The standard [1] indicates that when subchannelisation is employed, the SS maintains the same transmitted power density unless the maximum power level is reached. Consequently, when the number of active subchannels allocated to an uplink user is reduced, the transmitted power is reduced proportionally, without additional power control messages. When the number of subchannels is increased the total transmitted power is also increased proportionally. The transmitted power level must not exceed the maximum levels dictated by signal integrity considerations and regulatory requirements. The subchannelisation can then represent transmitted power decreases and, equivalently, capacity gains.

The 802.16e amendment defined an optional downlink subchannelisation zone in the OFDM PHY downlink subframe. Uplink subchannels are partly reused.

5.4 Subcarrier Permutations in WiMAX OFDMA PHY

Distributing the subcarriers over the subchannels is a very open problem with many parameters to consider: mobility, AAS support, different optimisation criterions, etc. The 802.16 standard and its amendment 802.16e provide full details for the many subcarriers permutations defined. In this section, we briefly describe the subcarriers permutations defined in Mobile WiMAX OFDMA PHY and detail one of these permutations.

5.4.1 The Main Permutation Modes in OFDMA

Subtracting the guard subcarriers and the DC subcarrier from N_{FFT} gives the set of 'used' subcarriers N_{used}. For both the uplink and downlink, these subcarriers are allocated as pilot subcarriers and data subcarriers according to one or another of the defined OFDMA permutation modes.

Two families of distribution modes can be distinguished:

- Diversity (or distributed) permutations. The subcarriers are distributed pseudo-randomly. This family includes: FUSC (Full Usage of the SubChannels) and PUSC (Partial Usage of the SubChannels), OPUSC (Optional PUSC), OFUSC (Optional FUSC) and TUSC (Tile Usage of SubChannels). The main advantages of distributed permutations are frequency diversity and intercell interference averaging. Diversity permutations minimise the probability of using the same subcarrier in adjacent sectors or cells. On the other hand, channel estimation is not easy as the subcarriers are distributed over the available bandwidth.
- Contiguous (or adjacent) permutations. These consider a group of adjacent subcarriers. This family includes the AMC (Adaptive Modulation and Coding) mode. This type of permutation leaves the door open for the choice of the best-conditions part of the bandwidth. Channel estimation is easier as the subcarriers are adjacent.

Mandatory permutation modes of the presently defined mobile WiMAX profiles are:

- for the downlink: PUSC, FUSC and AMC;
- for the uplink: PUSC and AMC.

5.4.2 Some OFDMA PHY Definitions

5.4.2.1 Subchannels and Pilot Subcarriers

A subchannel is the minimum transmission unit in an OFDMA symbol. Each of the permutation modes of OFDMA has its definition for a subchannel. There is also a difference between allocation of the data and pilot subcarriers in the subchannels between the different possible permutation modes:

Table 5.5 Slot definition

Permutation mode and communication way	Slot definition
Downlink FUSC; downlink OFUSC	1 subchannel × 1 OFDMA symbol
Downlink PUSC	1 subchannel × 2 OFDMA symbol
Uplink PUSC, uplink additional PUSC, downlink TUSC1 and TUSC1	1 subchannel × 3 OFDMA symbol
AMC (uplink and downlink)	1 subchannel × (1, 2 or 3) OFDMA symbol

- For (downlink) FUSC and downlink PUSC, the pilot tones are allocated first. What remains are data subcarriers, which are divided into subchannels that are used exclusively for data.
- For uplink PUSC, the set of used subcarriers is first partitioned into subchannels and then the pilot subcarriers are allocated from within each subchannel.

Thus, in the FUSC mode, there is one set of common pilot subcarriers, while in the uplink PUSC mode, each subchannel contains its own set of pilot subcarriers. For the downlink PUSC mode, there is one set of common pilot subcarriers for each major group including a set of subchannels (see below).

5.4.2.2 Slot and Burst (Data Region)

A slot in the OFDMA PHY has both a time and subchannel dimension. A slot is the minimum possible data allocation unit in the 802.16 standard. The definition of an OFDMA slot depends on the OFDMA symbol structure, which varies for uplink and downlink, for FUSC and PUSC, and for the distributed subcarrier permutations and the adjacent subcarrier permutation. See Table 5.5 for the different possibilities.

In OFDMA, a data region (or burst) is a two-dimensional allocation of a group of slots, i.e. a group contiguous subchannels, in a group of contiguous OFDMA symbols (see Figure 5.13 and the end of the PUSC section below for an example).

5.4.2.3 Segment

A segment is a subdivision of the set of available subchannels, used for deploying one instance of the MAC.

5.4.2.4 Permutation Zone

A permutation zone is a number of contiguous OFDMA symbols, in the downlink frame or the uplink frame, that use the same permutation mode. A downlink frame or an uplink frame may contain more than one permutation zones (see Figure 5.14), providing great malleability for designers.

5.4.3 PUSC Permutation Mode

The global principle of PUSC (Partial Usage of SubChannels) is the following. The symbol is first divided into subsets called clusters (downlink) or tiles (uplink). Pilots and data carriers

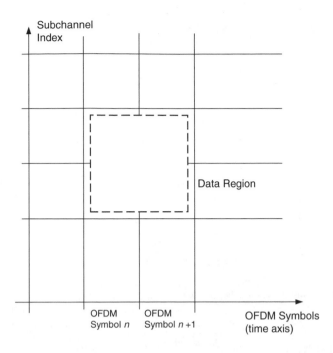

Figure 5.13 Example of the data region that defines the OFDMA burst allocation

are allocated within each subset. This allows partial frequency diversity. Some main MAC messages and some PHY subframe fields are transmitted in the PUSC mode: FCH, DL-MAP and UL-MAP (see Chapter 9 for these messages). Downlink PUSC subchannel allocation will now be detailed, which is illustrated by an example.

The global principle of downlink PUSC cluster and subcarrier allocation is illustrated in Figure 5.15. Considering, for example, a 1024-FFT OFDMA symbol, the number of guard subcarriers + DC carrier is (in the case of 1024 FFT) $92 + 91 + 1 = 184$. Therefore, the number of pilot and data carriers to be distributed is $1024 - 184 = 840$. The parameters of this numerical example are given in Table 5.6.

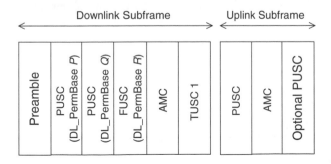

Figure 5.14 Example of different permutation zones in uplink and downlink frames

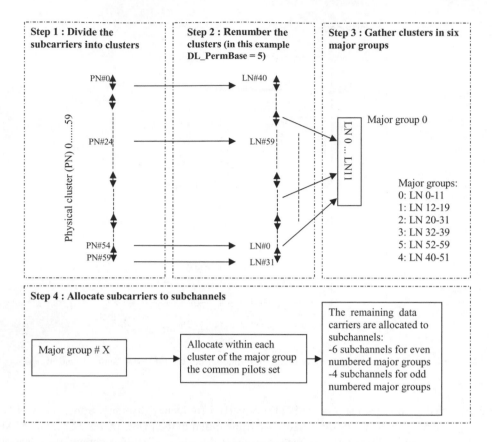

Figure 5.15 Illustration of the downlink PUSC Cluster and subcarrier allocation

5.4.3.1 Allocation Steps

Step 1. Divide the Subcarriers into Clusters

After removing the guard and DC subcarriers, the 840 (pilot and data) subcarriers are divided into 60 clusters of 14 adjacent subcarriers each ($14 \times 60 = 840$) (see Figure 5.16). We here mention that a PUSC cluster has nothing to see with a cluster of cells (see Chapter 4). The Physical Cluster number is between 0 and 59. Pilot subcarriers are placed within each cluster depending on the parity of the OFDMA symbols, as shown in Figure 5.17.

Step 2. Renumber the Clusters

The clusters are renumbered with Logical Numbers (LNs). The cluster LN is also between 0 and 59. In order to renumber the clusters, the *DL_PermBase* parameter is used.

Table 5.6 Numerical parameters of the downlink PUSC example

Parameter	FFT Size	BW	G	N	Pilot + data subcarriers	N_{FFT}	f_s	Δf
Value	1024	10 MHz	1/8	28/25	840	1024	11.2 MHz	10.9375 kHz

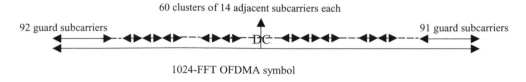

Figure 5.16 Cluster allocation

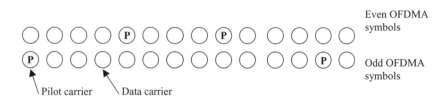

Figure 5.17 Cluster structure. (Based on Reference [2].)

DL_PermBase is an integer ranging from 0 to 31, which can be indicated by DL_MAP for PUSC zones (see Chapter 9).

The clusters are renumbered to LN clusters using the following formula (denoted Formula (0) in the following):

In the case of the first downlink zone (containing the FCH and DL-MAP), or *Use All SC Indicator = 0* in *STC_DL_Zone_IE* (see Chapter 9 for *STC_DL_Zone_IE*):

> **Cluster Logical Number = Renumbering sequence (Cluster Physical Number)** (0)
> else: **Cluster Logical Number**
> **= Renumbering sequence(((Cluster Physical Number)+13*DL_PermBase) mod**
> **Nclusters))**

where the *Renumbering sequence(j)* is the *jth* entry of the following vector:
[6, 48, 37, 21, 31, 40, 42, 56, 32, 47, 30, 33, 54, 18, 10, 15, 50, 51, 58, 46, 23, 45, 16, 57, 39, 35, 7, 55, 25, 59, 53, 11, 22, 38, 28, 19, 17, 3, 27, 12, 29, 26, 5, 41, 49, 44, 9, 8, 1, 13, 36, 14, 43, 2, 20, 24, 52, 4, 34, 0]

It should be remembered that, for 1024-FFT, *Nclusters = 60*, so the above vector has 60 elements.

Step 3. Gather Clusters in Six Major Groups
The renumbered clusters are then gathered in six major groups, using the LN, as shown in Table 5.7.

Step 4. Allocate Subcarriers to Subchannels
In the downlink PUSC the number of subchannels per OFDMA symbol is 30, numbered from 0 to 29. A subchannel is made of 24 data subcarriers, which represents the data subcarriers of two clusters. It can be verified that: $30 \times 24 = 720$ data subcarriers (720 data subcarriers + 30×4 pilot subcarriers $= 840$ subcarriers). For the downlink PUSC, each major group is used separately in order to have a number of subchannels; i.e. one subchannel does not have

Table 5.7   Downlink PUSC clusters major groups
(1024-FFT OFDMA)

Group	Cluster index
0	LN 0-11
1	LN 12-19
2	LN 20-31
3	LN 32-39
4	LN 40-51
5	LN 52-59

subcarriers in more than one major group. In addition, all the subcarriers of one subchannel belong to the same OFDMA symbol.

The pilot and data subcarrier allocations to subchannels are done as follows. The pilot subcarriers are allocated first within each cluster, placed as shown in Figure 5.17. In the downlink PUSC, there is one set of common pilot subcarriers in each major group. The remaining data subcarriers are first renumbered from 0 to 143 or 95 depending on the parity of the major group. Then the subcarriers are allocated within each subchannel using the following formula:

$$subcarrier(k,s) = N_{subchannels} * n_k + \{p_s[\, n_k \bmod N_{subchannels}] + DL\_PermBase\} \bmod N_{subchannels} \tag{5.1}$$

where $N_{subchannels}$ is the number of subchannels in the partitioned major group, equal to 4 or 6, depending on the parity of the major group; $subcarrier(k,s)$ is the subcarrier index of subcarrier k, varying between 0 and 23, in subchannel s, whose value ranges between 0 and 143 or 95 depending on the parity of the major group; s is the subchannel index varying between 0 and 29, and so

$$n_k = (k + 13s) \bmod N_{subcarriers}, \tag{5.2}$$

where $N_{subcarriers}$ is the number of data subcarriers allocated to a subchannel in each OFDMA symbol ($= 24$ in this case); $p_s[j]$ is the series obtained by rotating the basic permutation sequence cyclically to the left s times, which is given in the following: in the case of an odd numbered major group the basic permutation is *PermutationBase6* (3,2,0,4,5,1), while for an even numbered major group it is *PermutationBase4* (3,0,2,1).

For even numbered major groups, the 12 clusters contain the data subcarriers of 6 subchannels:

6 × 24 = 144 data subcarriers;
144 + 6 × 4 = 168 (data and pilot) subcarriers.

For odd numbered major groups, the 8 clusters contain the data subcarriers of 4 subchannels:

4 × 24 = 96 data subcarriers;
96 + 4 × 4 = 112 (data and pilot) subcarriers.

Table 5.8 Correspondence between subchannels and
major groups. (Based on Reference [2].)

Major group (subchannel group)	Subchannel range
0	0–5
1	6–9
2	10–15
3	16–19
4	20–25
5	26–29

The correspondance between subchannels and major groups is given in Table 5.8 (for 1024-FFT OFDMA). A numerical example of the downlink PUSC allocation is proposed below.

5.4.3.2 Numerical Example

Based on comprehension of the IEEE 802.16 standard, a numerical example is proposed. A start is made with step 4, the previous steps having fixed values. The aim is to find the 24 physical (data) subcarriers of subchannel 16 of the downlink PUSC. It is assumed that $DL\_PermBase = 5$ (indicated in the DL-MAP MAC Management Message) and that the OFDMA symbol considered is odd numbered.

Subchannel 16 is in major group 3, as shown in Table 5.8. Therefore, *basic permutation sequence* = (3,0,2,1), $N_{subcarriers} = 24$ (this is the case for all subchannels) and $N_{subchannels} = 4$ (odd numbered major group). In major group 3, the correspondence between the Logical Number (LN) and the original Physical Number (PN) is obtained by applying the equation of step 2, using the LN and its position in the renumbering sequence. Thus the correspondence is as shown in Table 5.9.

Table 5.10 depicts n_k and the physical subcarrier index corresponding to the each subcarrier k in subchannel s (= 16). For each subcarrier, the LN cluster of this subcarrier major group is used in order to find the physical subcarrier index (Table 5.9 is also used). For the pilot set for major group 3, using Table 5.9 values and the principle of Figure 5.16 gives the physical indices of each cluster pilot subcarriers. These indices are proposed in Table 5.11.

Table 5.9 Original cluster numbering (major group 3)

Cluster LN	Logical subcarrier index	Cluster PN formula (0)	Cluster physical subcarrier index
32	0–13	3	42–55
33	14–27	6	84–97
34	28–41	53	742–755
35	42–55	20	280–293
36	56–69	45	630–643
37	70–83	57	798–811
38	84–97	28	392–405
39	98–111	19	266–279

Table 5.10   Subcarrier allocation

Logical subcarrier index (k) in the considered subchannel ($s = 16$)	n_k (formula (5.2))	Logical subcarrier index in the major group (formula (5.1)) and corresponding cluster LN (using Table 5.9)		Physical subcarrier index (using Table 5.9 and Figure 5.17)
		Subcarrier	Cluster LN	
0	16	67	37	805
1	17	69	37	807
2	18	76	38	396
3	19	78	38	398
4	20	83	38	403
5	21	85	39	267
6	22	92	39	274
7	23	94	39	276
8	0	3	32	45
9	1	5	32	47
10	2	12	32	55
11	3	14	33	86
12	4	19	33	91
13	5	21	33	93
14	6	28	34	746
15	7	30	34	749
16	8	35	34	753
17	9	37	35	281
18	10	44	35	288
19	11	46	35	290
20	12	51	36	633
21	13	53	36	635
22	14	60	36	643
23	15	62	37	800

Table 5.11   Pilot subcarrier physical index

Cluster PN	Pilot subcarrier physical index
3	42 and 54
6	84 and 96
53	742 and 754
20	280 and 292
45	630 and 642
57	798 and 810
28	392 and 404
19	266 and 278

Table 5.12 Instantaneous data rate of one subchannel. (unit: kb/s)

	BPSK 1/2	QPSK 1/2	QPSK 3/4	16-QAM 1/2	16-QAM 3/4	64-QAM 2/3	64-QAM 3/4
Instantaneous data rate	116.6	233.3	349.8	466.5	699.75	932.9	1049

5.4.3.3 One Subchannel Instantaneous Data Rate

The data rate corresponding to one slot (equal to one subchannel over two slots) will now be computed. One OFDMA symbol duration is $102.9\,\mu s$ (see the numerical example parameters in Table 5.6 in Section 5.4.3 and Section 5.2.4 computations). A slot contains 48 subcarriers and then 48 modulation symbols. The instantaneous data rate of one subchannel can then be computed. The obtained values (no repetition) are given in Table 5.12.

5.4.3.4 Use of the Subchannels to Allocate Bandwidth

A slot is, by definition, in the downlink PUSC, made of 48 subcarriers:

1 subchannel $\times$ 2 OFDMA symbols = 24 subcarriers/OFDMA symbol $\times$ 2 OFDMA symbols
= 48 subcarriers (or 48 modulation symbols)

The data region information indicated in a DL-MAP MAC Management Message for an (OFDMA) downlink user is:

- OFDMA symbol offset. The offset of the OFDMA symbol in which the burst starts, measured in OFDMA symbols from the beginning of the downlink frame in which the DL-MAP is transmitted.
- Subchannel offset. The lowest index OFDMA subchannel used for carrying the burst, starting from subchannel 0.
- Number of OFDMA symbols. The number of OFDMA symbols that are used (fully or partially) to carry the downlink PHY burst. The value of the field is a multiple of the slot length in symbols.
- Number of subchannels. The number of subchannels with subsequent indexes used to carry the burst.

The parameter *DL_PermBase* is indicated in the zone switch IE (Information Element) in a DL-MAP, indicating a given permutation zone.

5.4.3.5 Uplink PUSC

For uplink PUSC, subchannels are built based on Tiles (see Figure 5.18). An uplink PUSC slot is made of one subchannel over three OFDMA symbols.

5.4.4 FUSC Permutation Mode

The global principle of FUSC (Full Usage of SubChannels) is close to PUSC. The difference is that there is no cluster (or tile) partitioning of subcarriers before subchannel allocation. Each subchannel subcarrier can be any where in the bandwidth.

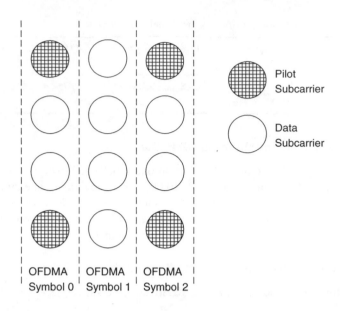

Figure 5.18   Uplink PUSC Tile is made of 12 subcarriers. (Based on Reference [1].)

In FUSC, the number of pilot and data carriers to be distributed is different from PUSC. The number of guard subcarriers + the DC carrier (in the case of 1024 FFT) is $87 + 86 + 1 = 174$. Therefore, the number of pilot and data carriers to be distributed is $1024 - 174 = 850$.

There are two constant pilot sets and two variable pilot sets (depending on the OFDMA symbol parity). Each segment uses both sets of variable/constant pilot sets.

For the FUSC mode, a 1024-FFT OFDMA symbol is considered. This symbol is divided into 16 subchannels of 48 subcarriers each, thus using all of the $16 \times 48 = 768$ data subcarriers. The data subcarriers are first divided into groups of contiguous subcarriers. Then each subchannel is constructed using one subcarrier of each group, as shown in Figure 5.19.

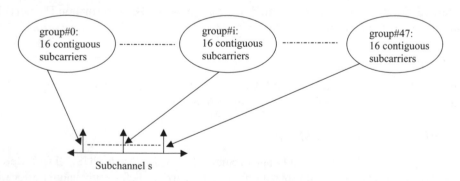

Figure 5.19   Illustration of a FUSC subchannel

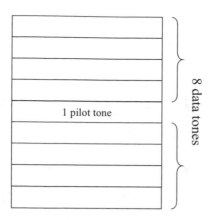

Figure 5.20 Bin structure

5.4.5 AMC Permutation Mode

In the AMC (Adaptive Modulation and Coding) mode, a subchannel is made of adjacent sub-carriers. The basic allocation unit for downlink and uplink AMCs is the bin, which is a set of nine adjacent (or contiguous) subcarriers within an OFDMA symbol, as shown in Figure 5.20. Contrary to distributed subchannel allocation techniques (such as PUSC), AMC pilot and data subcarriers are assigned fixed positions. This allows easy support of the AAS (Adaptive Antennas System).

An AMC slot is made of six bins or, equivalently, 48 data subcarriers and 6 pilot subcarriers. A group of four consecutive rows of bins in an OFDMA symbol is called a physical band. For 1024-FFT, the number of physical bands is then 24 ($24 \times 4 \times 9 = 864$ data and pilot subcarriers to be distributed). A grouping of physical AMC bands is a logical band. The maximum number of logical bands is specified in the Format configuration IE and can be a maximum of 24. For example, if the available physical bands are distributed over three logical bands, for 1024-FFT, a logical band is made of eight physical bands. The global principle of AMC subcarrier allocation is illustrated in Figure 5.21.

There are four types of AMC slots (48 subcarriers), which are combinations of six contiguous bins:

- Type 1 (default type). A slot consists of six consecutive bins within one OFDMA symbol.
- Type 2. A slot consists of 2 bins × 3 OFDMA symbols. This type is the only mandatory one according to presently defined Mobile WiMAX profiles.
- Type 3. A slot consists of 3 bins × 2 OFDMA symbols.
- Type 4. A slot consists of 1 bin × 6 OFDMA symbols.

The type is referred to as $N \times M$, where N is the number of bins and M the number of symbols.

5.4.5.1 AMC (or Regular AMC) and Band AMC

AMC allocations can be made by two mechanisms:

- Regular AMC (or, simply, AMC) is allocated by a subchannel index reference using DL-MAP and UL-MAP.

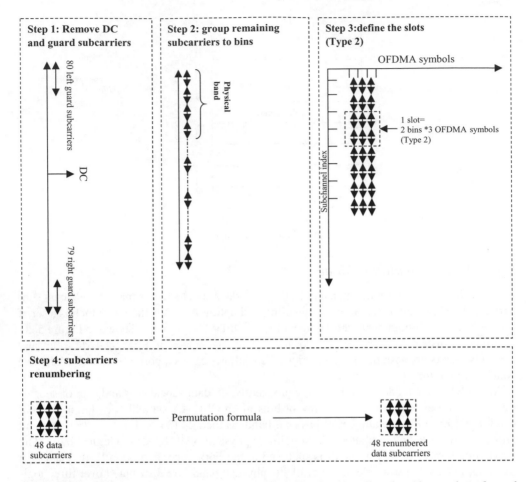

Figure 5.21   Illustration of the global principle of AMC subcarrier allocation (the number of guard subcarriers is for 1024-FFT)

- Band AMC is part of the HARQ map included with the normal map in the downlink or uplink zone. For Band AMC, the six contiguous bins of a Band AMC slot must be in the same logical band.

5.4.6 TUSC Permutation Mode

The IEEE 802.16e amendment defined two new optional permutation modes for the downlink: TUSC1 (Tile Usage of SubChannels) and TUSC2. TUSC1 and TUSC2 are dedicated to the AAS zone in an FDMA frame. The general principle is close to the uplink PUSC. TUSC1 and TUSC2 are not mandatory in presently defined mobile WiMAX profiles.

6

The Physical Layer of WiMAX

6.1 The 802.16 Physical Transmission Chains

The modulation and OFDM transmission aspects, described in the previous chapter, are the major building blocks of the WiMAX PHYsical Layer. In this chapter, some elements of the transmission chains of WiMAX are described for both OFDM and OFDMA PHYs.

6.1.1 The Global Chains

The PHY transmission chains of OFDM and OFDMA are illustrated in Figures 6.1 and 6.2. The blocks are the same with the small difference that OFDMA PHY includes a repetition block. The modulation is one of the four digital modulations described in the previous chapter: BPSK, QPSK, 16-QAM or 64-QAM. The modulated symbols are then transmitted on the OFDM orthogonal subcarriers. In the following, WiMAX channel coding building blocks are described.

The building blocks of channel coding are described in Section 6.2. A possible FEC code is the Turbo Code. Turbo Code theory and the basic elements of its use in WiMAX can be found in Section 6.3. The Transmission Convergence Sublayer (TCS), which can be applied in OFDM PHY, is described in Section 6.4. Finally, the burst profiles of OFDM and OFDMA PHY, an important building block of IEEE 802.16 MAC layer, are described in Section 6.5.

6.2 Channel Coding

The radio link is a quickly varying link, often suffering from great interference. Channel coding, whose main tasks are to prevent and to correct the transmission errors of wireless systems, must have a very good performance in order to maintain high data rates. The 802.16 channel coding chain is composed of three steps: randomiser, Forward Error Correction (FEC) and interleaving. They are applied in this order at transmission. The corresponding operations at the receiver are applied in reverse order. Error detection is realised with HCS and CRC (see Chapter 8).

WiMAX: Technology for Broadband Wireless Access   Loutfi Nuaymi
© 2007 John Wiley & Sons, Ltd

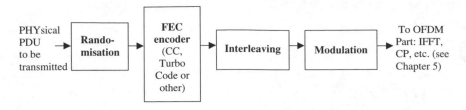

Figure 6.1 OFDM PHY transmission chain

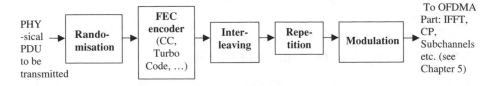

Figure 6.2 OFDMA PHY transmission chain

6.2.1 Randomisation

Randomisation introduces protection through information-theoretic uncertainty, avoiding long sequences of consecutive ones or consecutive zeros. It is also useful for avoiding non-centred data sequenes. Data randomisation is performed on each downlink and uplink burst of data. If the amount of data to transmit does not fit exactly the amount of data allocated, padding of 0×FF ('ones' only) is added to the end of the transmission block. The Pseudo-Random Binary Sequence (PRBS) generator used for randomisation is shown in Figure 6.3. Each data byte to be transmitted enters sequentially into the randomiser, with the Most Significant Byte (MSB) first. Preambles are not randomised. The randomiser sequence is applied only to information bits.

The shift-register of the randomiser is initialised for each new burst allocation. For OFDM PHY, on the downlink, the randomiser is reinitialised at the start of each frame with the sequence: 1 0 0 1 0 1 0 1 0 0 0 0 0 0 0 0. The randomiser is not reset at the start of burst 1. At the start of subsequent bursts (starting from burst 2), the randomiser is initialised with the vector

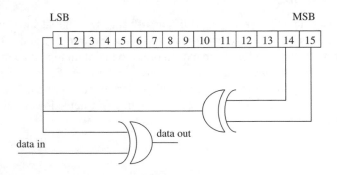

Figure 6.3 PRBS generator used for data randomisation in OFDM and OFDMA PHY. (From IEEE Std 802.16-2004 [1]. Copyright IEEE 2004, IEEE. All rights reserved.)

shown in Figure 6.4. This PRBS generates a Pseudo-Noise (PN) sequence of length $2^{15} - 1$. The frame number used for initialisation refers to the frame in which the downlink burst is transmitted. BSID is the BS identity and DIUC the burst profile indicator (see Chapter 9). For other cases (uplink, OFDMA), the details can be found in the standard. The bits issued from the randomiser are then applied to the FEC encoder.

6.2.2 Forward Error Correction (FEC) Codes

For OFDM PHY, the FEC encodings are:

- Concatenated Reed–Solomon Convolutional Code (RS-CC). This code is mandatory on both the uplink and downlink. It consists of the concatenation of a Reed–Solomon outer code and a rate-compatible convolutional inner code (see below).
- Convolutional Turbo Codes (CTC) (optional).
- Block Turbo Coding (BTC) (optional). For Turbo Coding, see Section 6.3 below.

The most robust burst profile or, equivalently, the most robust coding mode must be used when requesting access to the network and in the FCH burst (see Chapter 9 for FCH burst). For OFDMA PHY, the FEC encodings are:

- (Tail-biting) Convolutional Code (CC). This code is mandatory according to the 802.16 standard. According to WiMAX profiles, only the Zero-Tailing Convolutional Code (ZT CC) is mandatory.
- Convolutional Turbo Codes (CTC). This code is optional according to the 802.16 standards [1,2]. Yet, according to mobile WiMAX profiles, the CTC is mandatory.
- Block Turbo Coding (BTC) (optional).
- Low Density Parity Check (LDPC) codes (optional).

RS-CC encoding will now be described. WiMAX Turbo coding, BTC and CTC, will be described in the following section.

6.2.2.1 RS-CC (Reed–Solomon Convolution Code)

For OFDM PHY, the RS-CC encoding is performed by first passing the data in block format through the RS encoder and then passing it through a convolutional encoder (see Figure 6.5).

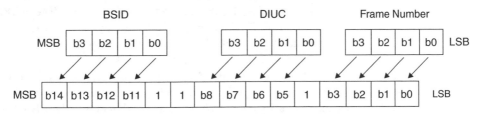

OFDM randomizer DL initialization vector

Figure 6.4 OFDM randomiser downlink initialisation vector for burst $2, \ldots, N$. (From IEEE Std 802.16-2004 [1]. Copyright IEEE 2004, IEEE. All rights reserved.)

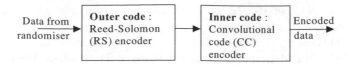

Figure 6.5   Illustration of the RS-CC encoder of OFDM PHY

Reed–Solomon codes are used in many communications systems and other applications. The RS error correction works by adding some redundant bits to a digital data sequence. This is done by oversampling a polynomial constructed from the uncoded data. The polynomial is evaluated at several points and then these values are sent (or recorded). By sampling the polynomial more often than needed, the receiver can recover the original polynomial in the presence of a relatively low number of errors.

A Reed–Solomon code is specified as $RS(N,K)$ with T-bit symbols. The data points are sent as encoded blocks. The total number of T-bit symbols in an encoded block is $N = 2^T-1$. Thus a Reed–Solomon code operating on 8-bit symbols has $N = 2^8-1 = 255$ symbols per coded block. The number K, $K < N$, of uncoded data symbols in the block is a design parameter. Then, the number of parity symbols added is $N - K$ symbols (of T-bits each). The RS decoder can correct up to $(N - K)/2$ symbols that contain an error in the encoded block.

The RS encoder of OFDM PHY is denoted as an $(N, K) = (255, 239)$ code, which is capable of correcting up to eight symbol errors per block. This Reed–Solomon encoding uses $GF(2^8)$, where GF is the Galois Field operator. The Reed–Solomon encoder and decoder require Galois field arithmethics. The following polynomials are used for the OFDM RS systematic code, an RS code that leaves the data unchanged before adding the parity bits:

Code generator polynomial: $g(x) = (x + \lambda^0)\,(x + \lambda^1)\,(x + \lambda^2)\cdots(x + \lambda^{2T-1})$, $\lambda = 02_{HEX}$;

Field generator polynomial: $p(x) = x^8 + x^4 + x^3 + x^2 + 1$.

The coding rate of the OFDM PHY RS encoder is then 239/255 (very close to one). The standard indicates that this code can be shortened and punctured to enable variable block sizes and variable error-correction capabilities.

The convolution code has an original coding rate of 1/2, as shown in Figure 6.6. The convolutional encoder is a zero-terminating convolutional encoder. A single 0×00 tail byte is appended to the end of each burst, needed for decoding algorithm normal operation. Puncturing patterns defined in the standard can be used to realise the following different code rates: 2/3, 3/4 and 5/6.

For OFDMA, the convolutional encoder is also the one shown in Figure 6.6. The HARQ procedure (described in Chapter 8), in its IR (Incremental Redundancy) variant, uses four different FEC blocks for each uncoded FEC block. This is realised using different puncture patterns. Each FEC block is identified by an SPID (SubPacket IDentifier).

The tail-biting convolutional code encoder of OFDMA (simply known as CC) works as follows: the convolutional encoder memories are initialised by the (six) last data bits of the FEC block being encoded (the packet data bits numbered $b_n - 5,\dots,b_n$). This OFDMA PHY convolutional encoder may employ the Zero-Tailing Convolutional Coding (ZT CC) technique. In this case, a single 0×00 tail byte is appended at the end of each burst. This tail byte is appended after randomisation.

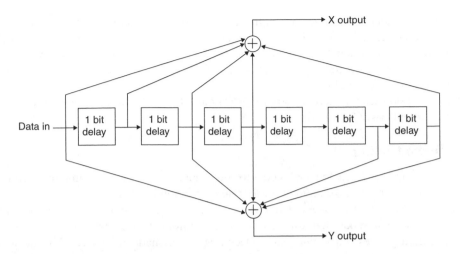

Figure 6.6 Convolutional encoder of rate 1/2. (From IEEE Std 802.16-2004 [1]. Copyright IEEE 2004, IEEE. All rights reserved.)

6.2.3 Interleaving

Interleaving is used to protect the transmission against long sequences of consecutive errors, which are very difficult to correct. These long sequences of error may affect a lot of bits in a row and can then cause many transmitted burst losses. Interleaving, by including some diversity, can facilitate error correction. The encoded data bits are interleaved by a block interleaver with a block size corresponding to the number of coded bits per allocated subchannels per OFDM symbol [1]. The interleaver is made of two steps:

- Distribute the coded bits over subcarriers. A first permutation ensures that adjacent coded bits are mapped on to nonadjacent subcarriers.
- The second permutation insures that adjacent coded bits are mapped alternately on to less or more significant bits of the constellation, thus avoiding long runs of bits of low reliability.

6.2.4 Repetition

Repetition was added by the 16e amendment for OFDMA PHY. The standard indicates that it can be used to increase the signal margin further over the modulation and FEC mechanisms.

In the case of repetition coding, $R = 2$, 4 or 6, the number of allocated slots (Ns) will be a whole multiple of the repetition factor R for the uplink. For the downlink, the number of the allocated slots (Ns) will be in the range of $R \times K, R \times K + (R - 1)$, where K is the number of required slots before applying the repetition scheme. For example, when the required number of slots before the repetition is 10 ($= K$) and the repetition of $R = 6$ is applied for the burst transmission, then the number of the allocated slots (Ns) for the burst can be from 60 slots to 65 slots.

The binary data that fits into a region that is repetition coded is reduced by a factor R compared to a nonrepeated region of the slots with the same size and FEC code type. After FEC and bit-interleaving, the data are segmented into slots, and each group of bits designated to fit in a slot is repeated R times to form R contiguous slots following the normal slot ordering that is used for data mapping.

This repetition scheme applies only to QPSK modulation. It can be applied in all coding schemes except HARQ with CTC.

6.3 Turbo Coding

Turbo codes are one of the few FEC codes to come close to the Shannon limit, the theoretical limit of the maximum information transfer rate over a noisy channel. The turbo codes were proposed by Berrou and Glavieux (from ENST Bretagne, France) in 1993. The main feature of turbo codes that make them different from the traditional FEC codes are the use of two error-correcting codes and an interleaver. Decoding is then made iteratively taking advantage of the two sources of information.

Data transmission is coded as follows (see Figure 6.7). Three blocks of bits are sent. The first block is the m-bit block of uncoded data. The second block is $n/2$ parity bits added in sequence for the payload data, computed using a convolutional code. The third subblock is another $n/2$ parity bits added in sequence for a known permutation of the payload data, also computed using a convolutional code. Hence, two different redundant blocks of parity bits are added to the sent payload. The complete block has $m + n$ bits of data with a code rate of $m/(m + n)$, as shown in the figure.

The data decoding process is the major innovation of turbo codes. The likelihood is used in order to take advantage of the differences between the two decoders. The turbo code inventors like to make the parallel with solving crosswords through both vertical and horizontal approaches.

Each of the two convolutional decoders generates an hypothesis, with derived likelihoods, for the m-bits sequence, called the a posteriori probability (APP). The hypothesis and the received sequence (recalculated) parity bits are compared and, if they differ, the decoder exchanges the derived likelihoods it has for each bit in the hypotheses. An iterative process is run until the two convolutional decoders come up with the same hypothesis for the m-bits sequence. The number of steps is usually of the order of 10.

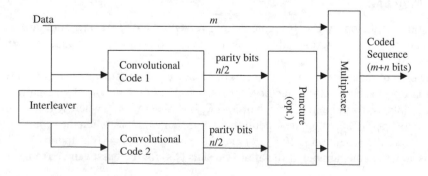

Figure 6.7   Turbo coded sequence generation

6.3.1 Convolutional Turbo Codes (CTC)

Different classes of turbo codes exist. Convolutional Turbo Codes (CTC) are defined as optional FEC for OFDM and OFDMA PHY. For OFDMA PHY, the CTC can be used for the support of the optional Hybrid ARQ (HARQ, see Chapter 8). According to mobile WiMAX profiles, the CTC is mandatory for OFDMA PHY. A brief overview of the CTC defined for OFDMA PHY is proposed here.

The CTC encoder, including its constituent encoder, is depicted in Figure 6.8. It uses a double binary Circular Recursive Systematic Convolutional Code. The bits of the data to be encoded are alternatively fed to A and B, starting with the MSB of the first byte being fed to A. The encoder is fed by blocks of k bits or N couples ($k = 2N$ bits). For all the frame sizes, k is a multiple of 8 and N is a multiple of 4. Further, N is limited to $8 \leq N/4 \leq 1024$.

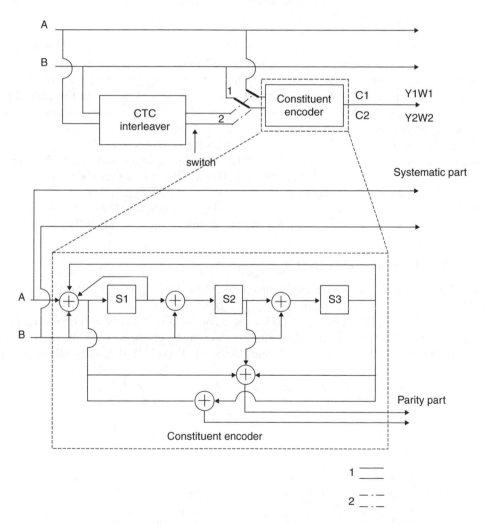

Figure 6.8   OFDMA PHY Convolutional Turbo Code (CTC) encoder. (From IEEE Std 802.16-2004 [1]. Copyright IEEE 2004, IEEE. All rights reserved.)

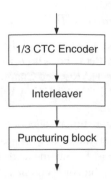

Figure 6.9   Block diagram of subpacket generation. (From IEEE Std 802.16-2004 [1]. Copyright IEEE 2004, IEEE. All rights reserved.)

   The encoding block size depends on the number of subchannels allocated and the modulation specified for the transmission. Concatenation of a number of subchannels must be performed in order to make larger blocks of coding where it is possible, with the limitation of not passing the largest block under the same coding rate. The concatenation rule should not be used when using HARQ. A table providing the number of subchannels concatenated as a function of the number of subchannels is given in the standard.
   Figure 6.9 shows a block diagram of CTC subpacket generation. The CTC encoded codeword with a coding rate of 1/3 goes through the interleaving block and puncturing is performed. FEC structures proposed in the standard [1] puncture the mother codeword to generate a subpacket with various coding rates: 1/2, 2/3, 3/4 and 5/6. The subpacket may also be used as HARQ packet generation (with different SPIDs). The length of the subpacket is chosen according to the needed coding rate, reflecting the channel condition (this is link adaptation).

6.3.2 Block Turbo Codes (BTC)

Block Turbo Codes (BTC) are defined as an optional FEC for OFDM and OFDMA PHY. The BTC is also optional in WiMAX profiles.
   For OFDM and OFDMA PHY, the BTC is based on the product of two simple component codes, which are binary extended Hamming codes or parity check codes. The codes are not the same for the two PHYs. BTC component codes of OFDM are shown in Table 6.1. The

Table 6.1   BTC component codes of OFDM PHY. (From IEEE Std 802.16-2004 [1]. Copyright IEEE 2004, IEEE. All rights reserved)

Component code (n,k)	Code type
(64,57)	Extended Hamming code
(32,26)	Extended Hamming code
(16,11)	Extended Hamming code
(64,63)	Parity check code
(32,31)	Parity check code
(16,15)	Parity check code
(8,7)	Parity check code

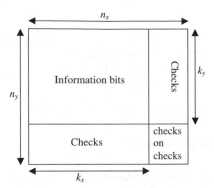

Figure 6.10 BTC and shortened BTC structure. (From IEEE Std 802.16-2004 [1]. Copyright IEEE 2004, IEEE. All rights reserved.)

component codes are used in a two-dimensional matrix form, which is depicted in Figure 6.10. The k_x information bits in the rows are encoded into n_x bits by using the component block (n_x, k_x) code specified in the standards for the respective composite code. After encoding the rows, the columns are encoded using a block code (n_y, k_y), where the check bits of the first code are also encoded. The overall block size of such a product code is $n = n_x\, n_y$, the total number of information bits $k = k_x\, k_y$ and the code rate is $R = R_x\, R_y$, where $R_i = k_i/n_i$, $i = x, y$. Data bit ordering for the composite BTC matrix is defined such that the first bit in the first row is the LSB (Least Significant Byte) and the last data bit in the last data row is the MSB.

To match a required packet size, BTCs may be shortened by removing symbols from the BTC array. In the two-dimensional case, rows, columns, or parts thereof, can be removed until the appropriate size is reached.

6.4 Transmission Convergence Sublayer (TCS)

The Transmission Convergence Sublayer (TCS) is defined in the OFDM PHY Layer and the Non-WiMAX SC PHY Layer. The TCS is located between the MAC and PHY Layers. If the TCS is enabled, the TCS converts MAC PDUs of variable size into proper-length FEC blocks, called TC PDU. An illustration of a TC PDU is shown in Figure 6.11. A pointer byte is added at the beginning of each TC PDU, as illustrated in the figure. This pointer indicates the header of the first MAC PDU.

The TCS is an optional mechanism for the OFDM PHY. It can be enabled on a pre-burst basis for both the uplink and downlink through the burst profile definitions in the uplink and downlink channel descriptor (UCD and DCD) messages respectively. The TCS_ENABLE parameter is coded as a TLV tuple in the DCD and UCD burst profile encodings (see Chapters 8 and 9 for TLV and UCD/DCD). At SS initialisation, the TCS capability is negotiated between the BS and SS through SBC-REQ/SBC-RSP MAC messages as an OFDM PHY specific parameter. The TCS is not included in the OFDMA PHY Layer.

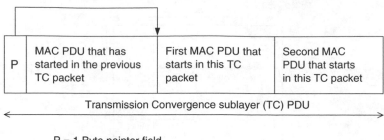

P = 1 Byte pointer field

Figure 6.11 Format of the downlink Transmission Convergence sublayer PDU. (From IEEE Std 802.16-2004 [1]. Copyright IEEE 2004, IEEE. All rights reserved.)

6.5 Burst Profile

The burst profile is a basic tool in the 802.16 standard MAC Layer. The burst profile allocation, which changes dynamically and possibly very fast, is about physical transmission. Here the parameters of the burst profiles of WiMAX are summarised. The burst profiles are used for the link adaptation procedure. The use of burst profiles and the link adaptation procedure will be seen in more detail in Chapters 9 and 10.

6.5.1 Downlink Burst Profile Parameters

The burst profile parameters of a downlink transmission for OFDM and OFDMA PHYsical layers are proposed in Table 6.2. The parameter called *FEC code* is in fact the Modulation and Coding Scheme (MCS). For OFDM PHY, there are 20 MCS combinations of modulation (BPSK, QPSK, 16-QAM or 64-QAM), coding (CC, RS-CC, CTC or BTC) and coding rate (1/2, 2/3, 3/4 and 5/6). The most frequency-use efficient (and then less robust) MCS

Table 6.2 Downlink burst profile parameters for OFDM and OFDMA PHYsical layers

Burst profile parameter	Description
Frequency (in kHz)	Downlink frequency
FEC code type	Modulation and Coding Scheme (MCS); there are 20 MCSs in OFDM PHY and 34 MCSs in OFDMA PHY (as updated in 802.16e)
DIUC mandatory exit threshold	The CINR at or below where this burst profile can no longer be used and where a change to a more robust (but also less frequency-use efficient) burst profile is required. Expressed in 0.25 dB units. See Chapter 9 for DIUC
DIUC minimum entry threshold	The minimum CINR required to start using this burst profile when changing from a more robust burst profile. Expressed in 0.25 dB units
TCS_ enable (OFDM PHY only)	Enables or disables TCS

Table 6.3 Uplink burst profile parameters for the OFDM PHYsical Layer

Burst profile parameter	Description
FEC type and modulation type	There are 20 MCSs in OFDM PHY
Focused contention power boost	The power boost in dB of focused contention carriers (see Chapter 10)
TCS_enable	Enables or disables TCS

Table 6.4 Uplink burst profile parameters for the OFDMA PHYsical Layer

Burst profile parameter	Description
FEC type and modulation type	There are 52 MCSs in OFDMA PHY
Ranging data ratio	Reducing factor, in units of 1 dB, between the power used for this burst and the power used for CDMA ranging (see Chapter 11); encoded as a signed integer

is 64-QAM (BTC) 5/6. For OFDMA PHY, there are 34 MCS combinations of modulation (BPSK, QPSK, 16-QAM or 64-QAM), coding (CC, ZT CC, CTC, BTC, CC with optional interleaver) and coding rate (1/2, 2/3, 3/4 and 5/6). The Downlink Interval Usage Code (DIUC) is the burst usage descriptor (see Chapter 9), which includes the burst profile.

6.5.2 Uplink Burst Profile Parameters

The burst profile parameters of an uplink transmission for an OFDM PHY and an OFDMA PHY are proposed in Tables 6.3 and 6.4 respectively.

6.5.3 MCS Link Adaptation

The choice between different burst profiles or, equivalently, between different MCSs is a powerful tool. Specifically, choosing the MCS most suitable for the state of the radio channel, at each instant, leads to an optimal (highest) average data rate. This is the so-called link

Table 6.5 Received SNR threshold assumptions [1], Table 266. (From IEEE Std 802.16-2004 [1]. Copyright IEEE 2004, IEEE. All rights reserved.)

Modulation	Coding rate	Receiver SNR threshold (dB)
BPSK	1/2	6.4
QPSK	1/2	9.4
QPSK	3/4	11.2
QAM-16	1/2	16.4
QAM-16	3/4	18.2
QAM-64	1/2	22.7
QAM-64	3/4	24.4

adaptation procedure. In the following chapters the MAC procedures that can be used for the implementation of link adaptation are described. The link adaptation algorithm in itself is not indicated in the 802.16 standard. It is left to the vendor or operator.

The order of magnitudes of SNR thresholds can be obtained from Table 6.5, proposed in the standard [1] for some test conditions. These SNR thresholds are for a BER, Bit-Error Rate, measured after the FEC, that is smaller than 10^{-6}.

Part Three

WiMAX Multiple Access (MAC Layer) and QoS Management

7

Convergence Sublayer (CS)

7.1 CS in 802.16 Protocol Architecture

The service-specific Convergence Sublayer (CS), often simply known as CS, is the top sub-layer of the MAC Layer in WiMAX/802.16 (Figure 7.1). The CS accepts higher-layer PDUs from the higher layers and transmits them to the MAC CPS where classical type MAC procedures are applied (see Chapter 8). Classifying and mapping the MSDUs into appropriate CIDs (Connection IDentifier) made by the CS are basic functions of the QoS mechanisms of WiMAX/802.16. Among other functions of the CS is the optional Payload Header Suppression (PHS), the process of suppressing repetitive parts of payload headers at the sender and restoring the headers at the receiver. The classification and mapping made by a QoS management module allow full advantage to be taken of the different PHYsical layer features presented in the two previous chapters of this book.

In the present version of the 802.16-2004 standard, two CS specifications are provided and described in Section 5 of standards [1] and [2]. The first CS specification is the ATM CS. The Asynchronous Transfer Mode (ATM) CS is a logical interface that associates different ATM services with the MAC CPS SAP. The ATM CS accepts ATM cells from the ATM layer, performs classification and, if provisioned, PHS. Then the ATM CS delivers CS PDUs to the appropriate MAC SAP.

The other available CS specification is the packet CS. The packet CS is used for the transport of all packet-based protocols such as the Internet Protocol (IP), IPv4, IPv6, Point-to-Point Protocol (PPP) and the IEEE standard 802.3 (Ethernet). Classification and, if provisioned, PHS are also defined for the packet CS.

The standard states that other CSs may be specified in the future. For the moment, no implementation of the ATM CS is planned, although it is detailed in the standard. In the rest of this chapter only the packet CS will be considered.

7.2 Connections and Service Flow

The CS provides any transformation or mapping of external network data received through the CS Service Access Point (SAP) into MAC SDUs received by the MAC Common Part Sublayer (CPS) through the MAC SAP (see Figure 7.1). This includes classifying external

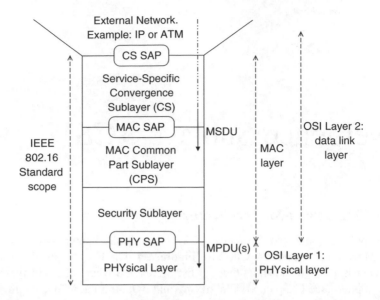

Figure 7.1   Protocol layers of the 802.16 BWA standard. (From IEEE Std 802.16-2004 [1]. Copyright IEEE 2004, IEEE. All rights reserved.)

network Service Data Units (SDUs) and associating them with the proper MAC Service Flow Identifier (SFID) and Connection Identifier (CID). Classification and mapping are then based on two 802.16 MAC layer fundamental concepts:

- Connection. A connection is a MAC Level connection between a BS and an SS (or MS) or inversely. It is a unidirectional mapping between a BS and an SS MAC peers for the purpose of transporting a service flow's traffic. A connection is only for one type of service (e.g. voice and email cannot be on the same MAC connection). A connection is identified by a CID (Connection IDentifier), an information coded on 16 bits.
- Service flow. A Service Flow (SF) is a MAC transport service that provides unidirectional transport of packets on the uplink or on the downlink. A service flow is identified by a 32-bit SFID (Service Flow IDentifier). The service flow defines the QoS parameters for the packets (PDUs) that are exchanged on the connection.

Figure 7.2 shows the relation between the SFID and CID. The relation between the two is the following: only admitted and active service flows (see the definitions below) are mapped to a CID, i.e. a 16-bit CID. In other terms:

- A SFID matches to zero (provisioned service flows) or to one CID (admitted or active service flow).

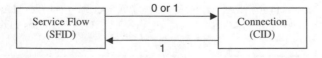

Figure 7.2   Correspondence between the CID and SFID

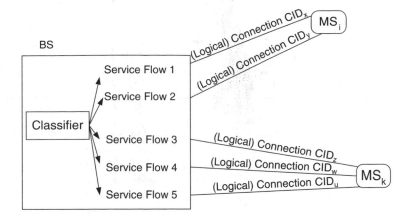

Figure 7.3   Illustration of service flows and connections

- A CID maps to a service flow identifier (SFID), which defines the QoS parameters of the service flow associated with that connection.

The definitions of connection and service flow in the 802.16 standard allow different classes of QoS to be found easily for a given element (SS or BS), with different levels of activation (see Figure 7.3). More details will now be given about connections (and CIDs) and service flows.

7.2.1 Connection Identifiers (CIDs)

A Connection IDentifier (CID) identifies a connection where every MAC SDU of a given communication service is mapped into. The CID is a 16-bit value that identifies a unidirectional connection between equivalent peers in the MAC layers of a BS and an SS.

All 802.16 traffic is carried on a connection. Then, the CID can be considered as a connection identifier even for nominally connectionless traffic like IP, since it serves as a pointer to destinations and context information [1]. The use of a 16-bit CID permits a total of 64K connections within each downlink and uplink channel. There are several CIDs defined in the standard (see Table 7.1). Some CIDs have a specific meaning. Some of the procedures introduced in this table, such as ranging, basic, primary and secondary management, AAS and others, will be introduced in different chapters later in this book.

Security associations (SAs) exist between keying material and CIDs, as described in Chapter 15.

7.2.2 Service Flows

A Service Flow (SF) is a MAC transport service that provides unidirectional transport of packets on the uplink or on the downlink. It is identified by a 32-bit SFID (Service Flow IDentifier).

A service flow is characterised by a set of QoS parameters. The QoS parameters include details of how the SS may request uplink bandwidth allocations and the expected behaviour

Table 7.1 CID ranges as defined in Reference [1]. Values are between 0000 (the 16 bits are equal to zero) and FFFF (the 16 bits are equal to one). It seems probable that the BS decides for a number m of CIDs for each of the basic and primary management connections that may be requested, i.e. a total of $2m$ connections. The CID value of basic, primary and secondary management connections for each SS are assigned in a ranging message (see Chapter 11). (From IEEE Std 802.16-2004 [1]. Copyright IEEE 2004, IEEE. All rights reserved.)

CID	Value	Description
Initial ranging	0×0000	Used by SS and BS during the initial ranging process
Basic CID	$0 \times 0001 - m$	Each SS has a basic CID and has a short delay. The same CID value is assigned to both the downlink and uplink connections
Primary management	$m + 1 - 2m$	The primary management connection is used to exchange longer, more delay-tolerant MAC management messages
Transport CIDs and secondary management CIDs	$2m + 1 - 0 \times FE9F$	Used for data transfer and for secondary management connection
Multicast CIDs	$0 \times FE9F - 0 \times FEFE$	For the downlink multicast service, the same value is assigned to all SSs on the same channel that participate in this connection
AAS initial ranging CID	$0 \times FEFF$	A BS supporting AAS (Advanced Antenna System) uses this CID when allocating an AAS ranging period (using AAS_Ranging_Allocation_IE)
Multicast polling CIDs	$0 \times FF00 - 0 \times FFF9$	An SS may be included in one or more multicast polling groups for the purposes of obtaining bandwidth via polling. These connections have no associated service flow
Normal mode multicast CID	$0 \times FFFA$	Used in DL-MAP to denote bursts for transmission of downlink broadcast information to normal mode SS
Sleep mode multicast CID	$0 \times FFFB$	Used in DL-MAP to denote bursts for transmission of downlink broadcast information to sleep mode SS. May also be used in MOB_TRF-IND messages
Idle mode multicast CID	$0 \times FFFC$	Used in DL-MAP to denote bursts for transmission of downlink broadcast information to idle mode SS. May also be used in MOB_PAG-ADV messages
Fragmentable broadcast CID	$0 \times FFFD$	Used by the BS for transmission of management broadcast information with fragmentation. The fragment subheader should use an 11-bit long FSN on this connection
Padding CID	$0 \times FFFE$	Used for transmission of padding information by the SS and BS
Broadcast CID	$0 \times FFFF$	Used for broadcast information that is transmitted on a downlink to all SSs

of the BS uplink scheduler. The main QoS parameters of the 802.16-2004 standard are given in Section 7.4 below. The service flow attributes are now given.

7.2.2.1 Service Flow Attributes

A service flow is partially characterised by the following attributes:

- Service Flow ID. An SFID is assigned to each existing service flow. The SFID serves as the identifier for the service flow in the network.
- CID. Mapping a CID to an SFID exists only when the connection has an admitted or active service flow (see below).
- ProvisionedQoSParamSet. This defines a QoS parameter set that is provisioned via means that the standard assumes to be outside of its scope. The standard states that this could be part of the network management system. For example, the service (or QoS) class name is an attribute of the ProvisionedQoSParamSet. There are five QoS classes, the fifth having been added by the 802.16e amendment.
- AdmittedQoSParamSet. This defines a set of QoS parameters for which the BS, and possibly the SS, are reserved resources. The principal resource to be reserved is bandwidth, but this also includes any other memory or time-based resource required to subsequently activate the flow.
- ActiveQoSParamSet. This defines a set of QoS parameters defining the service actually being provided to the service flow. Only an active service flow may forward packets. The activation state of the service flow is determined by the ActiveQoSParamSet. If the ActiveQoSParamSet is null, then the service flow is inactive.
- Authorisation module. This is a logical function within the BS that approves or denies every change to QoS parameters and classifiers associated with a service flow. As such, it defines an 'envelope' that limits the possible values of the AdmittedQoSParamSet and ActiveQoS-ParamSet (see Section 11.5.4 for more details about the authorisation module).

7.2.2.2 Types of Service Flow

The standard has defined three types of service flow:

- Provisioned service flows. This type of service flow is known via provisioning by, for example, the network management system. Its AdmittedQoSParamSet and ActiveQoSParamSet are both null.
- Admitted service flow. The standard supports a two-phase activation model that is often used in telephony applications. In the two-phase activation model, the resources for a call are first 'admitted' and then, once the end-to-end negotiation is completed, the resources are 'activated'.
- Active service flow. This type of service flow has resources committed by the BS for its ActiveQoSParamSet. Its ActiveQoSParamSet is non-null.

Each service flow class is associated with the corresponding QoSParametersSets. These three types of service flows can be seen as complementary. Figure 7.4 shows the possible transitions between these different service flows. A BS may choose to activate a provisioned service flow directly, or may choose to take the path to active service flows by passing the

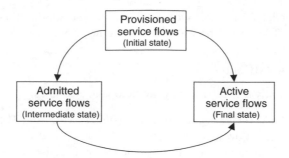

Figure 7.4 Possible transitions between service flows [1]

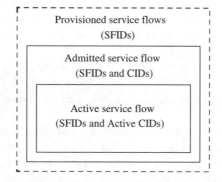

Figure 7.5 Model structure of the service flow types. (From IEEE Std 802.16-2004 [1]. Copyright IEEE 2004, IEEE. All rights reserved.)

admitted service flows. The model structure of these three service flow types is shown in Figure 7.5, taken from the standard.

More details about the activation of a service flow are given in Section 11.5. Having introduced the concepts of CID and SFID and their attributes, it is now possible to describe the process of classification and mapping made in the CS.

7.3 Classification and Mapping

As defined in the standard, classification is the process by which a MAC SDU is mapped on to a particular connection for transmission between MAC peers. The mapping process associates a MAC SDU with a connection, which also creates an association with the service flow characteristics of that connection. This process allows 802.16 BWA to deliver MAC SDUs with the appropriate QoS constraints.

Classification and mapping mechanisms exist in the uplink and downlink. In the case of a downlink transmission, the classifier will be present in the BS and in the case of an uplink transmission it is present in the SS.

A classifier is a set of matching criteria applied to each packet entering the WiMAX/802.16 network [1]. The set of matching criteria consists of some protocol-specific packet matching criteria (a destination IP address, for example), a classifier priority and a reference to a

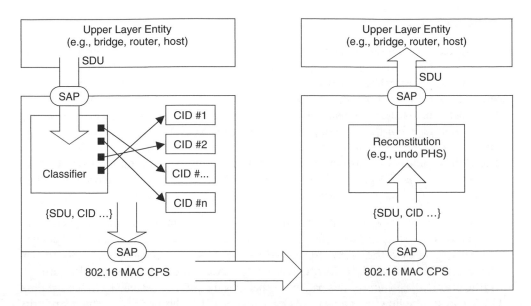

Figure 7.6 Classification and CID mapping. The principle is the same for both ways: BS to SS and SS to BS. (From IEEE Std 802.16-2004 [1]. Copyright IEEE 2004, IEEE. All rights reserved.)

CID. If a packet matches the specified packet matching criteria, it is then delivered to the SAP for delivery on the connection defined by the CID. The service flow characteristics of the connection provide the QoS for that packet. The classification mechanism is shown in Figure 7.6.

The MAC of 802.16-2004 is connection-oriented, where the connections are virtual. For the purposes of mapping simultaneous different services and associating varying levels of QoS, all data communications are in the context of a connection. Service flows may be provisioned when an SS is installed in the system. Shortly after SS registration, connections are associated with these service flows (it should be remembered that there is one connection per service flow) to provide a reference for the process of bandwidth request. Additionally, new connections may be established when an SS service needs change. A connection defines both the mapping between peer convergence processes that utilise the MAC and a service flow. The service flow defines the QoS parameters for the PDUs that are exchanged on the connection.

In short, the MAC CS Layer classifies each application. A QoS class is assigned. This classification is an important process because each BS may serve a relatively large number of users transmitting different applications. This classification allows a good link adaptation, because it will allocate the needed resources for each application. Consequently, the QoS differentiation, e.g. between an email and a voice transmission, is very easy to implement.

A Dynamic Service Addition (DSA) MAC management message allows the creation of a new service flow. MAC management messages are introduced in Chapter 8. The use of a DSA message for the creation of a new service flow is detailed in Chapter 11. The attributes of QoS are now given.

7.4 CS and QoS

During the creation of a service flow, the CS specification that the connection being set up will use is defined. Possible choices of CS specification are No CS, Packet IPv4, Packet IPv6, Packet 802.3/Ethernet, Packet 802.1Q VLAN, Packet IPv4 over 802.3/Ethernet, Packet IPv6 over 802.3/Ethernet, Packet IPv4 over 802.1Q VLAN, Packet IPv6 over 802.1Q VLAN and ATM.

The first feature of the QoS management is to define transmission ordering and scheduling on the air interface. For this reason, packets crossing the MAC interface are associated with a service flow identified by the CID.

The QoS parameters are associated with uplink/downlink scheduling for a service flow. An exhaustive list of QoS parameters of a service flow can be found in Reference [1], Section 11.13. The main QoS parameters are shortly described in the following:

- Scheduling service type, also called the QoS class. The value of this parameter specifies the scheduling service that is enabled for the associated service flow. The four defined scheduling service types in the 802.16 standard are (see Section 11.4 for scheduling service types) BE (default), nrtPS, rtPS and UGS. A fifth scheduling service type was added with 16e: ertPS.
- Traffic priority. The value of this parameter specifies the priority assigned to a service flow. Given two service flows identical in all QoS parameters besides priority, the higher priority service flow should be given lower delay and higher buffering preference.
- Maximum sustained traffic rate. This parameter defines the peak information rate of the service. The rate is expressed in bits per second.
- Maximum traffic burst. This parameter defines the maximum burst size that is accommodated for the service.
- Minimum reserved traffic rate. This parameter specifies the minimum rate reserved for this service flow. The rate is expressed in bits per second and specifies the minimum amount of data to be transported on behalf of the service flow when averaged over time.
- Vendor-specific QoS parameters. This allows vendors to encode vendor-specific QoS parameters. The Vendor ID must be embedded inside vendor-specific QoS parameters.
- Tolerated jitter. This parameter defines the maximum delay variation (jitter) for the connection.
- Maximum latency. The value of this parameter specifies the maximum latency between the reception of a packet by the BS or SS on its network interface and the forwarding of the packet to its RF interface.
- Fixed-length versus variable-length SDU indicator. The value of this parameter specifies whether the SDUs on the service flow are of a fixed length or variable length. This parameter is used only if packing is on for the service flow.
- SDU size. The value of this parameter specifies the length of the SDU for a fixed-length SDU service flow.
- Request/transmission policy. The value of this parameter provides the capability to specify certain attributes for the associated service flow. These attributes include options for PDU formation and, for uplink service flows, restrictions on the types of bandwidth request options that may be used.

7.5 Payload Header Suppression (PHS)

The packets delivered to OSI model layer 2 may have very large headers, sometimes as long as 120 bytes. This is the case for some RTP/UDP/IPv6 packets (RTP, Real-Time Protocol, UDP, User Datagram Protocol). This is very often repetitive (redundant) information and so

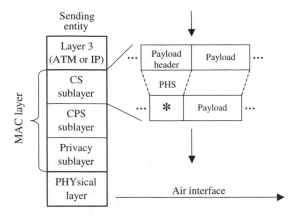

* Compressed payload header

Figure 7.7 Header suppression at the sending entity. Suppression of parts of the header leads to a compressed header. This allows the economy of a precious radio resource that would have been used for redundant information

should not be transmitted on a scarce resource such as a radio channel, which should be used for useful information. This process is known as header compression and decompression in 3G-cellular systems [13]. In the 802.16 standard, the PHS process suppresses repetitive (redundant) parts of the payload header in the MAC SDU of the higher layer. The receiving entity restores the suppressed parts. Implementation of the PHS capability is optional.

Figure 7.7 shows the PHS mechanism at the sending entity. Suppression of parts of the header leads to a compressed header. The receiver has to restore the header before properly using the received packet (Figure 7.8).

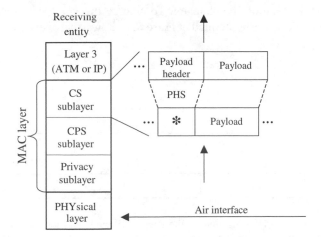

Figure 7.8 Header suppression mechanism at the receiving entity. The receiver has to restore the header before properly using the received packet

Table 7.2   Possible values of the PHS support field

Value	Description
0	No PHS support
1	ATM PHS
2	Packet PHS
3	ATM and Packet PHS

To indicate whether the PHS is present or not, the PHS support field is used. This parameter indicates the level of PHS support. The PHS support field is a field in some MAC management messages, Registration Request and Registration Response, which will be seen later in Section 11.6.1. Table 7.2 shows the possible values of the PHS support field. The default value is 0 (no PHS).

In the case of the ATM CS and in addition to the registration message, there is another possibility to indicate the PHS. In the DSA MAC management message, a field can implicitly signal the use of the PHS.

7.5.1 PHS Rules

PHS rules application and signalling are different for the two defined CS specifications: ATM CS and packet CS.

The ATM standard defines two modes: the VP-switched mode (Virtual Path) and the VC-switched mode (Virtual Channel). The same PHS operation is applied for the two modes; the only difference between them is the payload header size after suppression. When the PHS is turned off, no part of any ATM cell header including the Header Error Check (HEC) field is suppressed.

In the Packet CS mode, if the PHS is enabled at the MAC connection, each MAC SDU starts with a Payload Header Suppression Index (PHSI), an 8-bit field that references which Payload Header Suppression Field (PHSF) to suppress. Once a PHSF has been assigned to a PHSI, it cannot be changed. To change the value of a PHSF on a service flow, a new PHS rule must be defined. Figure 7.9 shows the operation of the PHS in a Packet CS.

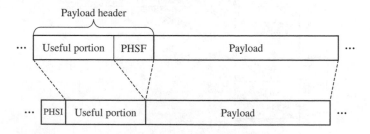

Figure 7.9   Illustration of PHS operation. In the Packet CS mode and for each MAC SDU, the PHSI (Payload Header Suppression Index) references the suppressed PHSF (Payload Header Suppression Field)

It is the responsibility of the higher-layer service entity to generate a PHS rule that uniquely identifies the suppressed header within the service flow. It is also the responsibility of the higher-layer service entity to guarantee that the byte strings that are being suppressed are constant from packet to packet for the duration of the active service flow.

As already mentioned, the sending entity uses the classifier to map packets into a service flow. The classifier uniquely maps packets to the PHS rule associated with this service flow. The receiving entity uses the CID and the PHSI to restore the PHSF.

The PHS has a Payload Header Suppression Valid (PHSV) option to verify the payload header before suppressing it. It also has a Payload Header Suppression Mask (PHSM) option to allow the choice of bytes that cannot be suppressed.

The PHS rule provides PHSF, PHSI, PHSM, PHSS (Payload Header Suppression Size) and PHSV.

7.5.2 PHS Rules Signalling

PHS rules are created with DSA or Dynamic Service Change (DSC) MAC management messages. The BS defines the PHSI when the PHS rule is created. PHS rules are deleted with the DSC or Dynamic Service Deletion (DSD) MAC management messages. When a classifier is deleted, any associated PHS rule is also deleted.

Figure 7.10 shows the use of a DSC message to signal the creation of a PHS rule and whether this rule is initiated by the BS or the SS. In this figure, the use of the following DSC MAC management messages are introduced:

- DSC-REQ. A DSC-REQ is sent by an SS or BS to dynamically change the parameters of an existing service flow. It can be used to create PHS rules.
- DSC-RSP. A DSC-RSP is generated in response to a received DSC-REQ.
- DSC-ACK. A DSC-ACK is generated in response to a received DSC-RSP.

The main DSC message parameters are SFID (only for DSC-RSP and DSC-ACK), CID, service class name, CS specification, etc. (see Chapter 11 for dynamic service management).

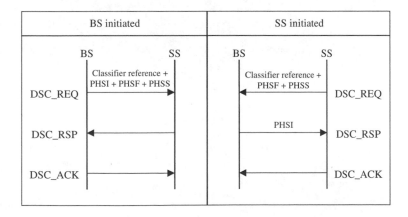

Figure 7.10 DSC MAC management message used for the signalling of a PHS rule. (From IEEE Std 802.16-2004 [1]. Copyright IEEE 2004, IEEE. All rights reserved.)

7.5.3 Header Compression in WiMAX

The PHS is a header suppression mechanism. There are also header compression algorithms that compress packet headers by other means than repetitive header suppression. The 802.16e amendment mentions that the Convergence Sublayer (CS) supports SDUs in two formats that facilitate robust compression of IP and higher-layer headers. These formats (corresponding to header compression algorithms) are:

- ROHC (RObust Header Compression), IETF RFC 3095 [14]. ROHC is also used in 3G UMTS cellular systems for header compression.
- ECRTP (Enhanced Compressed Real-Time Protocol) or Enhanced Compressed, RFC 3545 [15]. ECRTP is an evolution of CRTP, Compressed RTP, RFC 2508, defined for low data rate fixed line IP/UDP/RTP packet headers.

These two CS PDU formats defined in 802.16e are referred to as the IPheader-compression CS PDU format.

Compressed-IP-Header classifiers operate on the context fields of the ROHC and ECRTP compressed packets. 802.16e provides the service flow encodings that should be included in the CS packets for ROHC and ECRTP. 802.16e indicates that the compression function must not operate on the IEEE 802.3/Ethernet frame header so that the Ethernet frame header remains intact.

8

MAC Functions and MAC Frames

8.1 Introduction

In this chapter, some important 802.16 MAC functions and other MAC parameters are described. These functions and parameters will be used in the following three chapters, after which the MAC Layer of WiMAX will have been completely described. These functions may sometimes be complex. This is due to the two main objectives foreseen: to have a high bandwidth use efficiency and to provide full malleability for the QoS management.

First the MAC addresses and MAC frames format are described. Then, some MAC procedures such as fragmentation, packing and concatenation, ARQ and HARQ are revised. Basic, primary and secondary management connections are introduced. The TLV encoding often used in 802.16 is also introduced.

8.2 MAC Addresses and MAC Frames

8.2.1 MAC Addresses and Other Addresses

Each SS has a 48-bit universal MAC address, as defined in the standard [1]. This type of address is often known as the IEEE 802 MAC address. It uniquely defines the SS for all possible vendors and equipment types. It is used during the initial ranging process to establish the appropriate connections for an SS. It is also used as part of the authentication process by which the BS and SS each verify the identity of the other (see Chapter 11 for initial ranging and Chapter 15 for security). This is also the case in Mesh mode where each node has a unique IEEE 802 MAC address.

A 802.16 BS has a 48-bit Base Station ID (BSID). This is different from the MAC address of the BS. It includes a 24-bit operator indicator. The BSID can then be used for operator identification. It is used, for example, in the Downlink Channel Descriptor (DCD) MAC management message.

In the Mesh mode, another address in addition to the MAC address is used. When authorised to access, a candidate SS node receives a 16-bit Node IDentifier (Node ID) upon a request to an SS identified as the Mesh BS. The Node ID is the basis of node identification in the Mesh mode.

| MAC Header (6 bytes) | Payload (optional) | CRC (optional) (4 bytes) |

Figure 8.1   General format of a MAC frame or MAC PDU. (From IEEE Std 802.16-2004 [1]. Copyright IEEE 2004, IEEE. All rights reserved.)

8.2.2 MAC Frames

A MAC PDU is known as a MAC frame. It has the general format shown in Figure 8.1. Each MAC frame starts with a fixed-length MAC header. This header may be followed by the payload of the MAC PDU (MPDU). A MPDU may contain a CRC (Cyclic Redundancy Check). If present, the MPDU payload contains one or more of the following:

- zero or more subheaders included in the payload;
- zero or more MAC SDUs;
- fragment(s) of a MAC SDU.

The payload information may vary in length. Hence, a MAC frame length is a variable number of bytes. This format allows the MAC to tunnel various higher-layer traffic types without knowledge of the formats or bit patterns of those messages.

8.2.3 MAC Header Format

Two MAC header formats are defined in the standard:

- The Generic MAC Header (GMH). This is the header of MAC frames containing either MAC management messages or CS data. The CS data may be user data or other higher-layer management data. The generic MAC header frame is the only one used in the downlink.
- The MAC header without payload where two types are defined: Type I and Type II. For MAC frames with this type of header format, the MAC header is not followed by any MPDU payload and CRC. This frame name has been introduced by the 16e amendment. Previously, in 802.16-2004, the bandwidth request header was defined to request additional bandwidths (see Chapter 10). With 16e, the bandwidth request header becomes a specific case of MAC header formats without payload.

In the uplink, the single-bit Header Type (HT) field, at the beginning of the MAC header (see below), makes the distinction between the generic MAC header and the MAC header without payload formats: zero for the GMH and one for a header without payload.

8.2.3.1 Format of the Generic MAC Header

The Generic MAC frame header format is shown in Figure 8.2 and the fields of this header in Table 8.1. In this table, the LEN (length value) field, which is the length in bytes of the MAC frame (MAC PDU), is 11 bits long. This gives a maximal length of 2048 bytes for the MAC frame. The use of encryption parameters (EKS field) will be described in Chapter 15 where

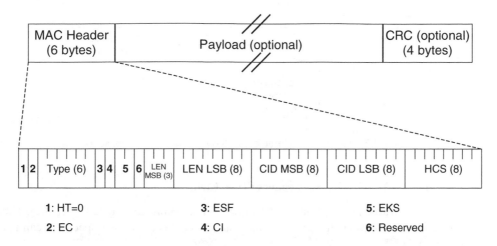

Figure 8.2 Header format of the Generic MAC frame [1]. The number in parentheses is the number of bits in the indicated field

WiMAX security is detailed. For both OFDM and OFDMA PHY layers, management messages must have a CRC, which is a 4-bytes field, computed as defined in IEEE Std 802.3 and appended to the payload of the MAC PDU. The CRC must be calculated after encryption, i.e. the CRC protects the GMH and the ciphered payload [1].

Table 8.1 Some fields of the MAC frame generic header. (From IEEE Std 802.16-2004 [1]. Copyright IEEE 2004, IEEE. All rights reserved.)

Name	Length (bits)	Description
HT	1	Header Type. Zero for generic MAC header
EC	1	Encryption Control: 0 = payload is not encrypted; 1 = payload is encrypted. For a MAC header without payload, this bit indicates whether it is Type I or II
Type	6	This field indicates the subheaders and special payload types present in the message payload (see below)
ESF	1	Extended Subheader Field. If ESF = 1, the extended subheader is present and follows the generic MAC header immediately (applicable in both the downlink and uplink)
CI	1	CRC Indicator: 1 = CRC is included (see MAC frame format); 0 = no CRC is included
EKS	1	Encryption Key Sequence. The index of the Traffic Encryption Key (TEK) and initialization vector used to encrypt the payload. Evidently, this field is only meaningful if the EC field is set to one
LEN	11	Length. The length in bytes of the MAC PDU including the MAC header and the CRC, if present
CID	16	Connection IDentifier (see Chapter 7)
HCS	8	Header Check Sequence. An 8-bit field used to detect errors in the header

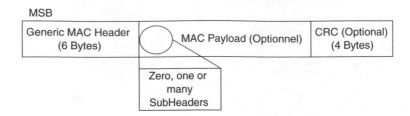

Figure 8.3 Generic MAC frame may have one or many subheaders

8.2.3.2 Generic MAC Header Subheaders (Type Field)

The generic MAC frame payload may start by one or more subheaders (see Figure 8.3). The type field in the generic MAC header is a 6-bit value, where each bit has a specific meaning, as shown in Table 8.2.

Fragmentation and packing are two MAC functions described in Section 8.3 where the use of the corresponding type field bit is given. ARQ (Automatic Repeat reQuest) is a handshake procedure of the data link layer where the receiver asks the transmitter to send again a block of data when errors are detected (ARQ is detailed further in this chapter in Section 8.7). If the ARQ feedback payload bit in the generic MAC header type field is set to 1, the ARQ feedback payload is transported. The extended type bit (bit 3) of the generic MAC header field indicates whether the present packing or fragmentation subheaders use ARQ-enabled connections. This bit is set to 1 if packing or fragmentation is applied on connections where ARQ is enabled.

8.2.3.3 Header Format Without Payload

The header without payload (Type I and II), only used in the uplink, has the same size as the generic MAC frame header, but the fields differ. The MAC header format without payload Type I is shown in Figure 8.4 and in Table 8.3. In the MAC header without payload, the header content field (19 bits) may have different uses. Type field encodings for the MAC signalling header Type I are given in Table 8.4 (from the 802.16e amendment). In this table the Bandwidth Request (BR) comprise dedicated bandwidth request frames. The BR field indicates the

Table 8.2 Encoding of the bits of the type field in the generic MAC header. (From IEEE Std 802.16-2004 [1]. Copyright IEEE 2004, IEEE. All rights reserved.)

Type bit	Value
5 (MSB)	Mesh subheader: 1 = present, 0 = absent
4	ARQ feedback payload: 1 = present, 0 = absent
3	Extended Type. Indicates whether the present packing or fragmentation subheaders is extended: 1 = extended, indicates connections where ARQ is enabled; 0 = not extended.
2	Fragmentation subheader: 1 = present, 0 = absent
1	Packing subheader: 1 = present, 0 = absent
0 (LSB)	Downlink: FAST-FEEDBACK allocation subheader; uplink: grant management subheader; 1 = present, 0 = absent

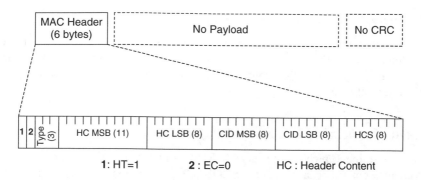

Figure 8.4   Header format without payload Type I. (Based on Reference [2].)

Table 8.3   Some fields of the MAC header without payload Type I. (Based on Reference [2].)

Name	Length (bits)	Description
HT	1	Header Type. One for the header without payload
EC	1	For a MAC header without payload, this bit indicates whether it is Type I or II
Type	3	Indicates the type of header without payload (see below)
Header content	19	Header content, function of the Type field value
CID	16	Connection IDentifier
HCS	8	Header Check Sequence (same as for the generic MAC header)

number of bytes requested; the CID indicates the connection for which the uplink bandwidth is requested. Aggregate and incremental BR types will be seen in Chapter 10. The SN report header is sent by the SS in the framework of the ARQ procedure.

In the MAC header format without payload Type II, the header is changed with regard to Type I. It is used for some feedbacks specific to OFDMA (MIMO, etc).

Table 8.4   Header format without payload Type I use. (Based on Reference [2].)

Type field (3 bits)	MAC header type (with HT/EC=0b10)
000	BR incremental
001	BR aggregate
010	PHY channel report
011	BR with UL Tx power report
100	Bandwidth request and CINR report
101	BR with UL sleep control
110	SN report
111	CQICH allocation request

8.2.3.4 Generic Frames: Transport or Management Frames?

The payload can contain either a management message or transport data. Specific connections are defined as management connections (see Table 7.1). These connections carry only management messages. All other connections carry user data or secondary (upper layer) MAC management data.

8.2.4 MAC Subheaders and Special Payloads

Use of the remaining Type bits of the generic MAC frame (see Table 8.2) are now described: Grant Management subheader, FAST_FEEDBACK_Allocation and Mesh subheader. The use of the corresponding subheaders is detailed.

Bandwidth requirements are not uniquely sent with a header without payload Type I bandwidth request header frames. The Grant Management subheader, which can be present only in the uplink, is used by the SS to transmit bandwidth management needs to the BS in a generic MAC header frame. This is then the so-called 'piggybacking request' as the data request takes place on a frame where data are also transmitted. The bandwidth request processes are described in Chapter 10, where details are given of the use of the Grant Management subheader (specifically in Section 10.2.2).

Fast feedback slots are slots individually allocated to SS for transmission of PHY-related information that requires a fast response from the SS. This allocation is done in a unicast manner through the FAST_FEEDBACK MAC subheader and signalled by Generic Header Type field bit 0. The FAST-FEEDBACK allocation is always the last per-PDU subheader. The FAST-FEEDBACK allocation subheader can be used only in the downlink transmission and with the OFDMA PHY specification (often with MIMO).

When authorised to a Mesh network, a candidate SS node receives a 16-bit Node IDentifier (Node ID) upon a request to the Mesh BS (see Section 3.6 for the Mesh BS). Node ID is the basis for identifying nodes during normal Mesh mode operation. The Mesh subheader contains a single information, the Node ID. If the Mesh subheader is indicated, it precedes all other subheaders.

8.3 Fragmentation, Packing and Concatenation

As in almost all other recent wireless systems, it may be interesting to fragment a MAC SDU in many MAC PDUs or, inversely, to pack more than one MSDU in many PDUs. The advantage of fragmentation is to lower the risk of losing a whole MSDU to the risk of losing part of it, a fragment. The inconvenient is to have more header information. This is interesting when the radio channel is relatively bad or packets too long. Conversely, packing allows less headers to be needed at the risk of losing all the packed packets. This is interesting when the radio channel is relatively good. Concatenation is the fact of transmitting many PDUs in a single transmission opportunity. Fragmentation, packing and concatenation are included in the 802.16 standard.

8.3.1 Fragmentation

Fragmentation is the process by which a MAC SDU is divided in two or more MAC PDUs. When the radio channel is relatively bad, this process allows efficient use of available

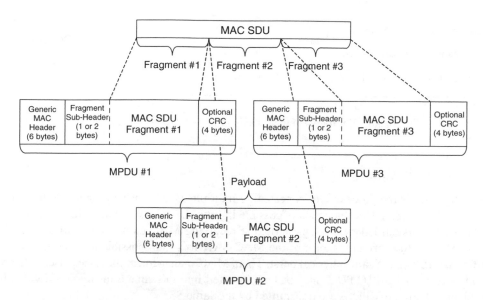

Figure 8.5 Illustration of the fragmentation of an MAC SDU giving three MAC PDUs (or MAC frames)

bandwidth while taking into account the QoS requirements of a connection service flow. The presence of fragmentation is indicated by bit 2 of the Type field (see Section 8.2) of a generic MAC frame. Usually, fragmentation concerns relatively long packets (such as IP packets). Fragmentation of a packet is shown in Figure 8.5.

The three MPDUs obtained in the example shown each contain a Fragment subheader. Thus bit 2 of the Type field in the generic MAC header will be set to 1 (see Section 8.2.3). The Fragment subheader will contain information such as if the fragment is the first, middle or last, etc.

The capabilities of fragmentation and reassembly are mandatory.

8.3.2 *Packing*

When packing is turned on for a connection, the MAC layer may pack multiple MAC SDUs into one single MAC PDU. When the radio channel is relatively good, this allows a better use of available resources. The transmitting side has the full decision of whether or not to pack a group of MAC SDUs in a single MAC PDU. The presence of packing is indicated by bit 1 of the Type field of the generic MAC frame (see Section 8.2.3).

Packing is especially efficient for relatively short packets. A packed packet is shown in Figure 8.6. The payload of the frame will contain many packing subheaders, and each one will be followed by its MAC SDU. The sum of packed headers is smaller than the sum of headers of normal SDUs. This is why packing saves bandwidth resources. On the other hand, if the packed PDU is lost, all component SDUs are lost (while possibly only one would have been lost if packing was not done).

If the ARQ mechanism is turned on, subheaders of fragmentation and packing are extended. For example, the subheaders of a packed packet are made of 3 bytes instead of 2.

The capability of unpacking is mandatory.

Generic MAC Header (6 bytes)	Packing Sub-Header (2 or 3 bytes)	MAC SDU	Packing Sub-Header (2 or 3 bytes)	MAC SDU	....	Optional CRC (4 bytes)

Figure 8.6 Illustration of the packing of MAC SDUs in one MAC PDU

8.3.3 Concatenation

Concatenation is the procedure of concatening multiple MAC PDUs into a single transmission (see Figure 8.7). Concatenation is possible in both the uplink and downlink. Since each MAC PDU is identified by a unique CID, the receiving MAC entity is able to present the MAC SDU to the correct instance of the MAC SAP. It is then possible to send MPDUs of different CIDs on the same physical burst. Then, MAC management messages, user data and bandwidth request MAC PDUs may be concatenated into the same transmission. Evidently, in the uplink all the MPDUs are transmitted by the same SS.

8.4 Basic, Primary and Secondary Management Connections

As already mentioned, connections are identified by a 16-bit CID. At SS initialisation, taking place at SS network entry, two pairs of management connections (uplink and downlink connections) are established between the SS and the BS, and a third pair of management connections may be optionally established. These three pairs of connections reflect the fact that there are three different levels of QoS for management traffic between an SS and the BS:

- The basic connection is used by the BS MAC and SS MAC to exchange short, time-urgent MAC management messages. This connection has a Basic CID (see Table 7.1).
- The primary management connection is used by the BS MAC and SS MAC to exchange longer, more delay-tolerant MAC management messages. This connection has a Primary Management CID (see Table 7.1). Table 8.5 and 8.6 list all of the 802.16-2004 and 802.16e MAC management messages. See Annex A for brief descriptions of each message. Tables 8.5 and 8.6, specify which MAC management messages are transferred on each of these two connections.

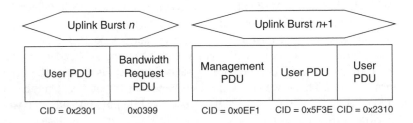

Figure 8.7 Illustration of the concatenation for an uplink burst transmission. (From IEEE Std 802.16-2004 [1]. Copyright IEEE 2004, IEEE. All rights reserved.)

Table 8.5 List of all 802. 16-2004 MAC management messages. See Annex A for brief descriptions of each message. (From IEEE Std 802. 16-2004 [1]. Copyright IEEE 2004, IEEE. All rights reserved.)

Type	Message name	Description	Connection
0	UCD	Uplink Channel Descriptor	Broadcast
1	DCD	Downlink Channel Descriptor	Broadcast
2	DL-MAP	Downlink Access Definition	Broadcast
3	UL-MAP	Uplink Access Definition	Broadcast
4	RNG-REQ	Ranging Request	Initial ranging or basic
5	RNG-RSP	Ranging Response	Initial ranging or basic
6	REG-REQ	Registration Request	Primary management
7	REG-RSP	Registration Response	Primary management
8	reserved		
9	PKM-REQ	Privacy Key Management Request	Primary management
10	PKM-RSP	Privacy Key Management Response	Primary management
11	DSA-REQ	Dynamic Service Addition Request	Primary management
12	DSA-RSP	Dynamic Service Addition Response	Primary management
13	DSA-ACK	Dynamic Service Addition Acknowledge	Primary management
14	DSC-REQ	Dynamic Service Change Request	Primary management
15	DSC-RSP	Dynamic Service Change Response	Primary management
16	DSC-ACK	Dynamic Service Addition Acknowledge	Primary management
17	DSD-REQ	Dynamic Service Deletion Request	Primary management
18	DSD-RSP	Dynamic Service Deletion Response	Primary management
19	reserved		
20	reserved		
21	MCA-REQ	Multicast Assignment Request	Primary management
22	MCA-RSP	Multicast Assignment Response	Primary management
23	DBPC-REQ	Downlink Burst Profile Change Request	Basic
24	DBPC-RSP	Downlink Burst Profile Change Response	Basic
25	RES-CMD	Reset Command	Basic
26	SBC-REQ	SS Basic Capability Request	Basic
27	SBC-RSP	SS Basic Capability Response	Basic
28	CLK-CMP	SS network Clock Comparison	Broadcast
29	DREG-CMD	De/Re-register Command	Basic
30	DSX-RVD	DSx Received Message	Primary management
31	TFTP-CPLT	Configuration File TFTP Complete Message	Primary management
32	TFTP-RSP	Configuration File TFTP Complete Response	Primary management
33	ARQ-Feedback	Standalone ARQ Feedback	Basic
34	ARQ-Discard	ARQ Discard message	Basic
35	ARQ-Reset	ARQ Reset message	Basic
36	REP-REQ	Channel measurement Report Request	Basic
37	REP-RSP	Channel measurement Report Response	Basic
38	FPC	Fast Power Control	Broadcast
39	MSH-NCFG	Mesh Network Configuration	Broadcast
40	MSH-NENT	Mesh Network Entry	Basic
41	MSH-DSCH	Mesh Distributed Schedule	Broadcast

(continued overleaf)

Table 8.5 (*continued*)

Type	Message name	Description	Connection
42	MSH-CSCH	Mesh Centralised Schedule	Broadcast
43	MSH-CSCF	Mesh Centralised Schedule Configuration	Broadcast
44	AAS-FBCK-REQ	AAS Feedback Request	Basic
45	AAS-FBCK-RSP	AAS Feedback Response	Basic
46	AAS-Beam_Select	AAS Beam Select message	Basic
47	AAS-BEAM_REQ	AAS Beam Request message	Basic
48	AAS-BEAM_RSP	AAS Beam Response message	Basic
49	DREG-REQ	SS De-registration Request message	Basic
50–255	reserved		

- The secondary management connection is used by the BS and SS to transfer delay tolerant, standards-based messages. These standards are the Dynamic Host Configuration Protocol (DHCP), Trivial File Transfer Protocol (TFTP), Simple Network Management Protocol (SNMP), etc. The secondary management messages are carried in IP datagrams, as mentioned later in Chapter 11 (see also Section 5.2.6 of the standard [1] for IP CS PDU formats). Hence, secondary management messages are not MAC management messages. Use of the secondary management connection is required only for managed SSs.

Table 8.6 MAC management messages added by the 802.16e amendment. (From IEEE Std 802.16e-2005 [2]. Copyright IEEE 2006, IEEE. All rights reserved.)

Type	Message name	Description	Connection
50	MOB_SLP-REQ	SLeep REQuest	Basic
51	MOB_SLP-RSP	SLeep ReSPonse	Basic
52	MOB_TRF-IND	TRaffic INDication	Broadcast
53	MOB_NBR-ADV	Neighbour ADVertisement	Broadcast and primary management
54	MOB_SCN-REQ	SCanning interval allocation REQuest	Basic
55	MOB_SCN-RSP	SCanning interval allocation ReSPonse	Basic
56	MOB_BSHO-REQ	BS HO REQuest	Basic
57	MOB_MSHO-REQ	MS HO REQuest	Basic
58	MOB_BSHO-RSP	BS HO Response	Basic
59	MOB_HO-IND	HO INDication	Basic
60	MOB_SCN-REP	Scanning result REPort	Primary management
61	MOB_PAG-ADV	BS broadcast PAGing	Broadcast
62	MBS_MAP	MBS MAP	—
63	PMC_REQ	Power control Mode Change REQuest	Basic
64	PMC_RSP	Power control Mode Change Response	Basic
65	PRC-LT-CTRL	Set-up/tear-down of Long-Term MIMO precoding	Basic
66	MOB_ASC-REP	Association result REPort	Primary management
67–255	reserved		

Management message type (1 Byte)	Management Message Payload

Figure 8.8 General format of a MAC management message (payload of a MAC PDU)

An SS supports a Basic CID, a Primary Management CID and zero or more Transport CIDs. A managed SS also supports a Secondary Management CID. Then the minimum value of the number of uplink CIDs supported is three for managed SSs and two for unmanaged SSs.

The CIDs for these connections are assigned in the initial ranging process, where the three CID values are assigned. The same CID value is assigned to both members (uplink and downlink) of each connection pair. The initial ranging process is described in Chapter 11.

8.5 User Data and MAC Management Messages

A transport connection is a connection used to transport user data. MAC management messages are not carried on transport connections. A transport connection is identified by a transport connection identifier, a unique identifier taken from the CID address space that uniquely identifies the transport connection.

A set of MAC management messages is defined. These messages are carried in the payload of a MAC PDU starting with a generic MAC header. All MAC management messages begin with a management message Type field and may contain additional fields. This field is 1 byte long. The format of the MAC management message is given in Figure 8.8.

MAC management messages on the basic, broadcast and initial ranging connections can neither be fragmented nor packed. MAC management messages on the primary management connection and the secondary management connection may be packed and/or fragmented. For the SCa, OFDM and OFDMA PHY layers, management messages carried on the initial ranging, broadcast, basic and primary management connections must have a CRC field.

The list of 802.16-2004 MAC management messages and the encoding of their management message Type field are given in Table 8.5. The 802.16e amendment added some new messages, given in Table 8.6. The new messages related to mobility start with MOB. In Annex A, the different sets of MAC management messages and the descriptions of these messages are shown. Many of these messages will be used in the following chapters.

MAC management messages very often include TLV encoding. TLV encoding is introduced in the next section.

8.6 TLV Encoding in the 802.16 Standard

A TLV encoding consists of three fields (a tuple): Type, Length and Value. TLV is a formatting scheme that adds a tag to each transmitted parameter containing the parameter type and the length of the encoded parameter (the value). The type implicitly contains the encoding rules. TLV encoding is used for parameters in MAC management messages. It is also used for configuration, definition of parameters like software updates, hardware version, Vendor ID, DHCP, etc.

The length of the Type field is 1 byte. The lengths of the remaining fields is explained in the following.

If the length of the Value field is less than or equal to 127 bytes, then the length of the Length field is 1 byte, where the most significant bit is set to 0. The other 7 bits of the Length field are used to indicate the length of the Value field in bytes.

If the length of the Value field is more than 127 bytes, then the length of the Length field is one byte more than is needed to indicate the length of the Value field in bytes. The most significant bit is set to 1. The other 7 bits of the first byte of the Length field are used to indicate the number of additional bytes of the Length field (i.e. excluding this first byte). The remaining bytes (i.e. excluding the first byte) of the Length field are used to indicate the length of the Value field.

Disjoint sets of TLVs are made that correspond to each functional group. Each set of TLVs that are explicitly defined to be members of a compound TLV structure form an additional set. Unique Type values are assigned to the member TLV encodings of each set. Uniqueness of TLV Type values is then assured by identifying the IEEE 802.16 entities (MAC management messages and/or configuration file) that share references to specific TLV encodings.

8.6.1 TLV Encoding Sets

In Table 8.7, a brief description is given of TLV encoding sets in the 802.16 standard. For each encoding set, the section of the standard is given where details of this encoding can be found. For some TLV sets, the standard defines TLV encoding parameters for each PHY specification.

In this table, it can be verified that the Type values of common TLV encoding sets are unique (when compared to other sets). This is the only collection for which global uniqueness is guaranteed.

Annex B of this book provides a detailed example of TLV coding use in 802.16.

8.7 Automatic Repeat Request (ARQ)

The ARQ (Automatic Repeat reQuest) [16] is a control mechanism of data link layer where the receiver asks the transmitter to send again a block of data when errors are detected. The ARQ mechanism is based on acknowledgement (ACK) or nonacknowledgement (NACK) messages, transmitted by the receiver to the transmitter to indicate a good (ACK) or a bad (NACK) reception of the previous frames. A sliding window can be introduced to increase the transmission rate. Figure 8.9 shows the cumulative ARQ mechanism.

An ARQ block is a distinct unit of data that is carried on an ARQ-enabled connection. An ARQ block is assigned a sequence number (SN) or a Block Sequence Number (BSN) and is managed as a distinct entity by the ARQ state machines. The block size is a parameter negotiated during connection establishment.

A system supporting ARQ must then be able to receive and process the ARQ feedback messages. The ARQ feedback information can be sent as a standalone MAC management message (see Type 33 in Table 8.5) on the appropriate basic management connection or piggybacked on an existing connection. Piggybacked ARQ feedback is sent as follows: the ARQ feedback payload subheader, introduced in Section 8.2.3 (see Type 4 bit in the generic MAC frame header), can be used to send the ARQ ACK variants: cumulative, selective, selective

Table 8.7 Brief descriptions of TLV encoding sets in the 802.16 standard. Several Type values are common to different sets but no confusion is possible

Encodings set	Type	Description
Common encodings	143 → 149	Define parameters such as current transmit power, downlink/uplink service flow descriptor, HMAC (see Chapter 15) information, etc. Some of these parameters are used by the other TLV encoding sets. Section 11.1 of the standard
Configuration file encodings	1 → 7	Only for the configuration (Section 9 of the standard). Define parameters like software updates, hardware version, Vendor ID, etc. Section 11.2 of the standard
UCD management message encodings	1 → 5	Define uplink parameters such as the uplink burst profile that can be used (see Chapter 9). Section 11.3 of the standard
DCD management message encodings	1 → 17	Define downlink parameters such as the downlink burst profile that can be used (see Chapter 9). Section 11.4 of the standard
RNG-REQ management message encodings	1 → 4	Define Ranging Request parameters such as the requested downlink burst profile. Section 11.5 of the standard
RNG-RSP management message encodings	1 → 13	Define ranging response parameters. Example: Basic CID and Primary management CID are TLV RNG-REQ encoded parameters. Section 11.6 of the standard
REG-REQ/RSP management message encodings	1 → 17	Define Registration Request parameters such as CS capabilities, ARQ parameters, etc. (see Chapter 11). Section 11.7 of the standard
SBC-REQ/RSP management message encodings	1 → 4	Define SS Basic Capability Request parameters such as physical parameters supported and bandwidth allocation support (see Chapter 11). Section 11.8 of the standard
PKM-REQ/RSP management message encodings	6 → 27 except 14, 25 and 26	Define security-related parameters like SAID (Security Association IDentifier), SS certificate, etc. (see Chapter 15) Section 11.9 of the standard.
MCA-REQ management message encodings	1 → 6	Define Multicast Assignment Request parameters like Multicast CID, periodic allocation type, etc. Section 11.10 of the standard
REP-REQ management message encodings	1	Define parameters related to channel measurement report request. Section 11.11 of the standard
REP-RSP management message encodings	1 and 2	Define parameters related to channel measurement report which is the response to channel measurement report request. Section 11.12 of the standard.
Service flowmanagement encodings	1 → 28 except 4 and 27, 99 → 107 and 143	Define the parameters associated with uplink/downlink scheduling for a service flow like SFID, CID, etc. Section 11.13 of the standard

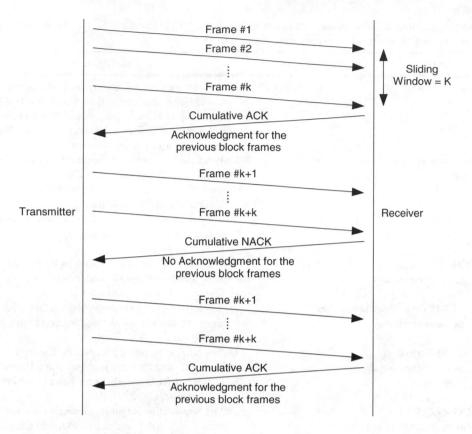

Figure 8.9   Illustration of the cumulative ARQ process

with cumulative, cumulative with block. When sent on an appropriate basic management con-nection, the ARQ feedback cannot be fragmented.

The ARQ is a MAC mechanism which is optional for implementation in the 802.16 standard. When implemented, the ARQ may be enabled on a per-connection basis. The per-connection ARQ is specified and negotiated during connection creation. A connection cannot have a mixture of ARQ and non-ARQ traffic.

8.7.1 ARQ Feedback Format

The Standalone ARQ Feedback message can be used (in addition to piggybacking ARQ) to signal any combination of different ARQ ACKs: cumulative, selective and selective with cumulative. Table 8.8 shows the ARQ Feedback Information Element (IE) used by the receiver of an ARQ block to signal positive or negative acknowledgments. The ACK map is a field where each bit indicates the status (received correctly or not) of the referred ARQ block.

If ACK Type $= 0 \times 1$ (cumulative ARQ), the BSN value indicates that its corresponding block and all blocks with lesser values within the transmission window have been success-fully received. Figure 8.9 represents the cumulative ACK ARQ mechanism.

Table 8.8   ARQ Feedback Information Element (IE) contents (the list is nonexhaustive). (From
IEEE Std 802.16-2004 [1]. Copyright IEEE 2004, IEEE. All rights reserved)

Field	Size	Notes
CID	16 bits	The ID of the connection being referenced
Last	1 bits	0 = more ARQ feedback IE in the list; 1 = last ARQ feedback IE in the list
ACK Type	2 bits	0×0 = Selective ACK entry, 0×1 = cumulative ACK entry, 0×2 = cumulative with selective ACK entry, 0×3 = cumulative ACK with a block sequence ACK entry
BSN (Block Sequence Number)	11 bits	The definition of this field is a function of ACK Type. For cumulative ACK, the BSN value indicates that its corresponding block and all blocks with lesser values within the transmission window have been successfully received
Number of ACK maps	2 bits	If ACK Type = 01, the field is reserved and set to 00. Otherwise this field indicates the number of ACK maps: 0×0 = 1, 0×1 = 2, 0×2 = 3 and 0×3 = 4
(For selective ACK types) one or more selective ACK Maps	16 bits per ACK map	Each bit set to one indicates that the corresponding ARQ block has been received without errors. The bit corresponding to the BSN value in the IE is the most significant bit of the first map entry

8.7.1.1 Selective ACK and Cumulative with Selective ACK

Each bit set to one in the selective ACK MAP indicates that the corresponding ARQ block has been received without errors. The bit corresponding to the BSN value in the ARQ Feedback Information IE is the most significant bit of the first map entry. The bits for succeeding block numbers are assigned left-to-right (MSB to LSB) within the map entry.

Cumulative with selective ACK associates cumulative and selective mechanisms: ACKnowledgement is made for a number of ARQ blocks.

8.7.1.2 ARQ and Packing or Fragmentation

The ARQ mechanism may be applied to fragmented MAC SDUs or to packed MAC PDUs. In this case, the Extended Type bit in the generic MAC header must be set to 1 (see Section 8.2).

8.7.2 Hybrid Automatic Repeat Request (HARQ) Mechanism

The Hybrid ARQ (HARQ) mechanism uses an error control code in addition to the retransmission scheme to ensure a more reliable transmission of data packets (relative to ARQ). The main difference between an ARQ scheme and an HARQ scheme is that in HARQ, subsequent retransmissions are combined with the previous erroneously received transmissions in order to improve reliability. HARQ parameters are specified and negotiated during the initialisation procedure. A burst cannot have a mixture of HARQ and non-HARQ traffic. The HARQ

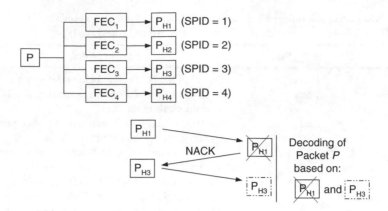

Figure 8.10   Incremental Redundancy (IR) HARQ

scheme is an optional part of the 802.16 standard MAC. HARQ may only be supported by the OFDMA PHYsical interface.

For the downlink HARQ, a fast ACK/NACK exchange is needed. Uplink slots ACK (UL ACK) in the OFDMA frame allow this fast feedback (see the OFDMA frame in Chapter 9). Two main variants of HARQ are supported:

- Incremental Redundancy (IR) for CTC and CC. The PHY layer encodes the HARQ packet generating several versions of encoded subpackets (see Figure 8.10). Each subpacket is uniquely identified by a SubPacket IDentifier (SPID). Four subpackets can be generated for a packet to be encoded. For each retransmission the coded block (the SPID) is different from the previously transmitted coded block.
- Chase Combining (CC) for all coding schemes. The retransmission is identical to the initial transmitted block. The PHY layer encodes the HARQ packet generating only one version of the encoded packet (no SPID is required).

An SS may support IR and an SS may support either CC or IR.

8.8 Scheduling and Link Adaptation

Scheduling will be described in Chapter 11. At this stage, some scheduling principles will be introduced that will be used before Chapter 11. Scheduling services are globally the data handling mechanisms allowing a fair distribution of resources between different WiMAX/802.16 users. Each connection is associated with a single data service and each data service is associated with a set of QoS parameters that quantify aspects of its behaviour, known as a QoS class.

Four classes of QoS were defined in the 802.16-2004 standard (and then in WiMAX):

- Unsolicited Grant Service (UGS);
- real-time Polling Service (rtPS);
- non-real-time Polling Service (nrtPS);
- Best Effort (BE).

A fifth one has been added with 802.16e: extended real-time Polling Service (ertPS) class.

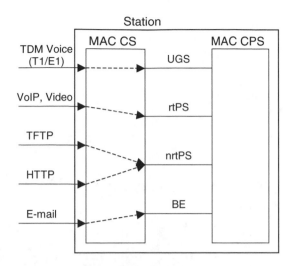

Figure 8.11   Scheduling mechanisms in a station (BS or SS). (From IEEE Std 802.16-2004 [1]. Copyright IEEE 2004, IEEE. All rights reserved.)

The purpose of scheduling is to allow every user, if possible, to have the suitable QoS required for his or her application. For example, a user sending an email does not require a real-time data stream, unlike another user having a Voice over IP (VoIP) application.

The main mechanism for providing QoS is to associate packets crossing the MAC interface into a service flow as identified by the CID. As already mentioned in the previous chapter, the MAC CS layer makes the classification of different user applications in these five classes of services. Once that operation is made, the role of the MAC CPS layer is to provide the connection establishment and maintenance between the two sides. Figure 8.11 shows an illustration of scheduling mechanisms in a station (BS or SS).

In the PMP mode, the BS controls both uplink and downlink scheduling. Uplink request/grant (see Chapter 9) scheduling is performed by the BS with the intent of providing each subordinate SS with a bandwidth for uplink transmissions or opportunities to request the bandwidth.

The link adaptation allows a fair performance for the different applications and a good optimisation of using the radio resources, realising the QoS required for the transmission of the data streams. The link adaptation is an adaptive modification of the burst profile, mainly modulation and channel coding types, that take place in the physical link to adapt the traffic to a new radio channel condition. If the CINR decreases, change is made to a robust modulation and coding to improve the performance (data throughput); otherwise a less robust profile is picked up. For more details on the link adaptation and burst profile transitions see Chapter 11.

9

Multiple Access and Burst Profile Description

9.1 Introduction

The aim of this chapter is to describe the multiple access of WiMAX/802.16. It will be seen that the mechanisms of multiple access and radio resource sharing are rather complex. It can be said that they are more complex than in other known wireless systems such as GSM, WiFi/IEEE 802.11 or even UMTS. Yet, globally, WiMAX multiple access is an extremely flexible F/TDMA (Frequency and Time Division Multiple Access).

The concept of a service flow on a connection is central to the operation of the MAC protocol. Service flows in the 802.16 standard provide a mechanism for QoS management in both the uplink and downlink. Service flows are integral to the bandwidth allocation process. In this process, an SS requests an uplink bandwidth on a per-connection basis (implicitly identifying the service flow). Bandwidth is granted by the BS to an SS in response to per-connection requests from the SS. WiMAX has been called a Demand Assigned Multiple Access (DAMA) system.

First, duplexing possibilities are described in Section 9.2. Physical frames are described in Section 9.3. WiMAX transmissions take place on totally dynamic bursts. The concept of multiple access is tightly related to burst profile. Frame contents are indicated in DL-MAP and UL-MAP messages, described in Section 9.4. The concept of a burst in the 802.16 standard and the way burst profiles are announced by the BS are detailed in Section 9.5. The specific case of the Mesh mode is tackled in Section 9.6.

9.2 Duplexing: Both FDD and TDD are Possible

The WiMAX/802.16 standard includes the two main duplexing techniques: Time Division Duplexing (TDD) and Frequency Division Duplexing (FDD). The choice of one duplexing technique or the other may affect certain PHY parameters as well as impact on the features that can be supported. Next, each of these duplexing techniques will be discussed.

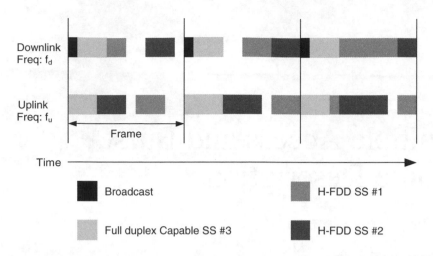

Figure 9.1   Illustration of different FDD mode operations: broadcast, full duplex and half duplex. Half duplex SSs as SS 1 and 2 in this figure can listen to the channel or (exclusively) send information. (From IEEE Std 802.16-2004 [1]. Copyright IEEE 2004, IEEE. All rights reserved.)

9.2.1 FDD Mode

In an FDD system, the uplink and downlink channels are located on separate frequencies. A fixed duration frame is used for both uplink and downlink transmissions. This facilitates the use of different modulation types. It also allows simultaneous use of both full-duplex SSs, which can transmit and receive simultaneously and, optionally, half-duplex SSs (H-FDD for Half-duplex Frequency Division Duplex), which cannot. A full-duplex SS is capable of continuously listening to the downlink channel, while a half-duplex SS can listen to the downlink channel only when it is not transmitting on the uplink channel. Figure 9.1 illustrates different cases of the FDD mode of operation.

When half-duplex SSs are used, the bandwidth controller does not allocate an uplink bandwidth for a half-duplex SS at the same time as the latter is expected to receive data on the downlink channel, including allowance for the propagation delay uplink/downlink transmission shift delays.

9.2.2 TDD Mode

In the case of TDD, the uplink and downlink transmissions share the same frequency but they take place at different times. A TDD frame (see Figures 9.2 and 9.3) has a fixed duration and contains one downlink and one uplink subframe. The frame is divided into an integer number of Physical Slots (PSs), which help to partition the bandwidth easily. For OFDM and OFDMA PHYsical layers, a PS is defined as the duration of four modulation symbols. The frame is not necessarily divided into two equal parts. The TDD framing is adaptive in that the bandwidth allocated to the downlink versus the uplink can change. The split between the uplink and downlink is a system parameter and the 802.16 standard states that it is controlled at higher layers within the system.

Mesh topology supports only TDD duplexing.

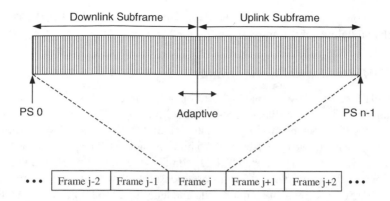

Figure 9.2 TDD frame: uplink and downlink transmissions share the same frequency but have different transmission times. (From IEEE Std 802.16-2004 [1]. Copyright IEEE 2004, IEEE. All rights reserved.)

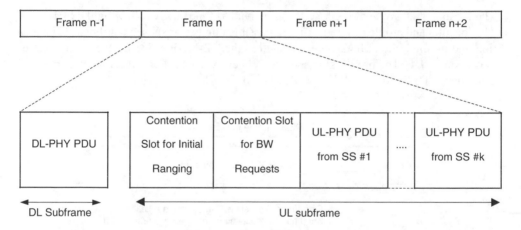

Figure 9.3 General format of a TDD frame (OFDM PHY). (Based on Reference [1].) In the FDD mode, the downlink subframe and uplink subframes are transmitted on two separate frequencies as the uplink frame and downlink frame. The contents are the same for FDD and TDD.

Comparing the two modes, a fixed duration frame is used for both uplink and downlink transmissions in FDD while the TDD distribution is adaptive. Therefore TDD duplexing is more suitable when data rates are asymmetrical (between the uplink and downlink), e.g. for an Internet transmission.

After settling the question of duplexing, many users have to share the bandwidth resource in each kind of transmission.

9.3 Transmission of Downlink and Uplink Subframes

Downlink and uplink transmissions coexist according to one of the two duplexing modes: TDD or FDD. They are sent through the downlink and uplink subframes. More specific information is now given about each of these two subframes for OFDM PHY. The structure of downlink and uplink subframes is the same for TDD and FDD.

9.3.1 OFDM PHY Downlink Subframe

An OFDM PHY downlink subframe consists of only one downlink PHY PDU, this PDU being possibly shared by more than one SS. A downlink PHY PDU starts with a long preamble, which allows PHY synchronisation for listening SSs. A listening SS synchronises to the downlink using the preamble (see Section 9.3.5 and Chapter 11). The preamble is followed by a Frame Control Header (FCH) burst. The FCH contains the Downlink Frame Prefix (DLFP) which specifies the burst profile and length of at least one downlink burst immediately following the FCH. Several downlink burst profile and lengths, up to four after the FCH, may be indicated in the DLFP. An HCS field occupies the last byte of the DLFP.

For OFDM PHY, the standard indicates that the DLFP is one OFDM symbol with the most robust modulation and coding scheme. The modulation and coding scheme can be considered to be BPSK with a coding rate of 1/2. In the DLFP, the following are specified:

- The location and profile of the first downlink burst (immediately following the FCH).
- The location and profile of the maximum possible number of subsequent bursts. The location and profile of other bursts are specified in the DL-MAP MAC management message (see Section 9.4). The profile(s) is specified either by a 4-bit Rate_ID (for the bursts indicated by the DLFP) or by DIUC (in DL-MAPs).

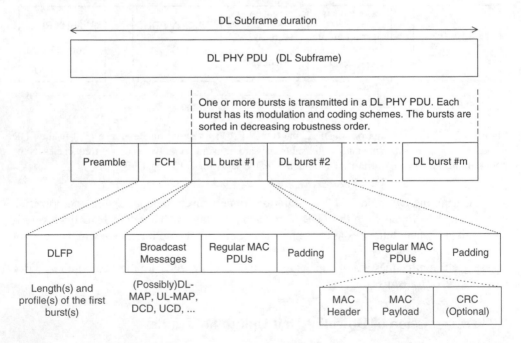

Figure 9.4 Details of the OFDM PHY downlink subframe. Each downlink burst may be sent to one (unicast) or more SSs (multicast or broadcast).

Each downlink burst may be sent to one (unicast) or more SSs (multicast or broadcast). A DL-MAP message (indicator of the downlink frame use, see below), if transmitted in the current frame (a case where no DL-MAP is needed: the DLFP indicates all the burst profiles of the downlink subframe), must be the first MAC PDU in the burst following the FCH. A UL-MAP message (indicator of the uplink frame use, see below) immediately follows either the DL-MAP message (if there is one) or the FCH. If UCD and DCD messages are transmitted in the frame, they immediately follow the DL-MAP and UL-MAP messages. The FCH is followed by one or many downlink bursts. The same burst profile can be used more than one time (this is a 16e update of 802.16-2004 which required each burst being transmitted with a different burst profile). These downlink bursts are transmitted in order of decreasing robustness of their burst profiles. The general format of a downlink subframe is shown in Figure 9.4. Burst profile indicators are DIUC and UIUC, described in the sequel.

9.3.2 OFDM PHY Uplink Subframe

Figure 9.5 represents the structure of an uplink subframe. An OFDM PHY uplink subframe consists of three global parts in this order:

- Contention slots allowing initial ranging. Via the Initial Ranging IE, the BS specifies an interval in which new stations may join the network (see Chapter 11 for the initial ranging procedure). Packets transmitted in this interval use the RNG-REQ (Ranging Request) MAC management message and are transmitted using a contention procedure as collision(s) may occur with other incoming SSs (see Chapter 10 for the contention procedures).

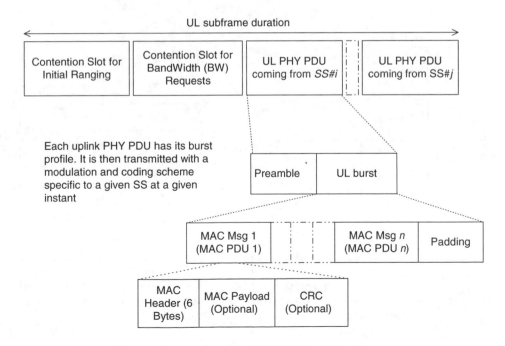

Figure 9.5 Details of the OFDM PHY uplink subframe.

- Contention slots allowing bandwidth requests. Via the Request IE, the BS specifies an up-link interval in which requests may be made for a bandwidth for uplink data transmission (see Chapter 10 for bandwidth request).
- One or many uplink PHY PDUs, each transmitted on a burst. Each of these PDUs is an uplink subframe transmitted from a different SS. A PDU may transmit an SS MAC messages.

9.3.3 OFDMA PHY Frame

For the OFDMA PHY Layer, the frame format is evidently different, taking into account that data mapping is made on two dimensions: time and subcarriers. Figure 9.6 shows an example of an OFDMA frame in the TDD mode. This figure includes nonmandatory OFDMA frame elements.

The transitions between modulations and coding take place on slot boundaries in the time domain (except in the AAS zone) and on subchannels within an OFDMA symbol in the frequency domain. The FCH is transmitted using the QPSK rate 1/2 with four repetitions using the mandatory coding scheme. Then, the FCH information is sent on four adjacent subchannels with successive logical subchannel numbers in a PUSC zone. The FCH contains the DLFP which specifies the length of the DL-MAP message that immediately follows the DLFP and the repetition coding used for the DL-MAP message.

The OFDMA frame may include multiple zones (such as PUSC, FUSC, PUSC with all subchannels, optional FUSC, AMC, TUSC1 and TUSC2). The transition between zones is indicated in the DL-MAP by the STC_DL_Zone or AAS_DL_IE. Both of these DIUCs are

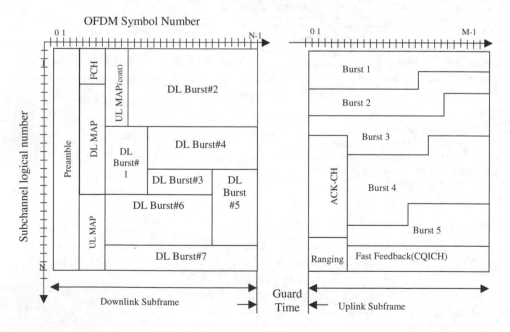

Figure 9.6 Example of an OFDMA frame in the TDD mode. (Based on References [2] and [10].)

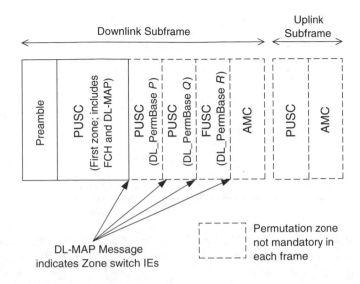

Figure 9.7   Illustration of the OFDMA frame with multiple zones. (Based on Reference [2].)

extended DIUC (=15) specific assignments. DL-MAP and UL-MAP allocations cannot span over multiple zones. Figure 9.7 shows an OFDMA frame with multiple zones. In the first PUSC zone of the downlink (first zone), the default renumbering sequence is used for cluster logical numbering.

The frame structure used for the uplink includes:

- Allocation for ranging. The uplink ranging subchannel is allocated for SSs for ranging (initial/periodic/handover ranging) and bandwidth requests.
- The fast feedback slot includes four bits of payload data, whose encoding may contain CINR measurements, handover operation messages, extended rtPS bandwidth request, etc. The BS may allocate a CQICH (Channel Quality Information CHannel) (also called a fast-feedback channel) using a CQICH_IE (CQICH_allocation_IE or CQICH_Control_IE) for periodic CINR reports. This uplink channel state information feedback is used for some handover and MIMO operations. The CQICH also exists for the downlink.
- Other optional signalling data allocations are handover-related subchannels, MIMO-related subchannels, HARQ UL subchannel, HARQ ACK subchannel, Power_control_IE, AAS_UL_IE, etc.
- Allocation for data transmission.

9.3.4 Frame Duration

Frame duration possible values are dependent on the PHYsical Layer. The frame duration values for the OFDM (WiMAX) PHY Layer are shown in Table 9.1 with the corresponding frame duration codes. For the OFDMA PHY Layer, a value is added to this list: 2 ms. For mobile WiMAX (OFDMA) system profiles only a 5 ms duration is mandatory.

Table 9.1 Frame duration possible values for OFDM
(WiMAX) PHY Interface (based on [1])

Frame duration code	Frame duration (ms)
0	2.5
1	4
2	5
3	8
4	10
5	12.5
6	20
7–255	*reserved*

The frame duration is decided by the BS. This value is transmitted in the DCD message on the frame duration code (on 8 bits), as seen in Section 9.5.2 and Annex B. For the two duplexing systems, the rule is the following:

- In an FDD system, the uplink and downlink channels are located on separate frequencies. A fixed duration frame is used for both uplink and downlink transmissions.
- In the case of TDD, the uplink and downlink transmissions occur at different (complementary) times while sharing the same frequency. A TDD frame contains one downlink and one uplink subframe. The general format of an OFDM PHY TDD frame is shown in Figure 9.3. For OFDMA PHY, the format is evidently different, taking into account the two dimensions: time and subcarriers (see Figure 9.6).

We now describe the preambles used in 802.16.

9.3.5 Preambles

A 802.16 preamble is a standard-defined sequence of symbols known by the receiver. The preamble is used by the PHYsical Layer for synchronisation and equalisation. The preamble must be taken into account for precise computation of a useful data rate.

For the OFDM PHY Layer, all preambles are structured as either one (short preamble) or two (long preamble) OFDM symbols. The OFDM symbols are defined by the values of the composing subcarriers. The Cyclic Prefix (CP) of those OFDM symbols has the same length as the CP of data OFDM symbols.

The long preamble is used in the following cases:

- the first preamble in the downlink PHY PDU;
- the initial ranging preamble;
- the AAS preamble.

The short preamble is used in the following cases:

- the first preamble in the uplink PHY PDU, when no subchannelisation is applied;
- in the downlink bursts that fall within the STC-encoded region, the preamble transmitted from both transmit antennas simultaneously;
- a burst preamble on the downlink bursts when indicated in the DL-MAP_IE.

In the case where the uplink allocation contains midambles, the midambles consist of one OFDM symbol and are identical to the preamble used with the allocation.

For the OFDMA PHY Layer, the preamble is a number of subcarriers.

9.4 Maps of Multiple Access: DL-MAP and UL-MAP

The broadcasted DL-MAP and UL-MAP MAC management messages define the access to the downlink and uplink information respectively. The DL-MAP is a MAC management message that defines burst start times on the downlink. Equivalently, the UL-MAP is a set of information that defines the entire (uplink) access for all SSs during a scheduling interval. Then DL-MAP and UL-MAP are directories, broadcasted by the BS, of downlink and uplink frames. Figure 9.8 shows an example of DL-MAP and UL-MAP use in the FDD mode.

For OFDM and OFDMA (or both WiMAX) PHY layers, access grants of DL-MAP and UL-MAP are in units of symbols and (for OFDMA) subchannels. Timing information in the DL-MAP and UL-MAP is relative. The following time instants are used as references for timing information for each of these two timings:

- DL-MAP: the start of the first symbol (including the preamble if present) of the frame in which the message was transmitted;
- UL-MAP: the start of the first symbol (including the preamble if present) of the frame in which the message was transmitted plus the value of the Allocation Start Time (whose value is given in the UL-MAP message, see below).

Information in the DL-MAP is about the current frame (the frame in which the DL-MAP message is sent). Information carried in the UL-MAP concerns a time interval starting at the Allocation Start Time measured from the beginning of the current frame and ending after the last specified allocation. Therefore, two possibilities exist concerning which frame is concerned with UL-MAP (differentiated by the Allocation Start Time field in the UL-MAP):

- UL-MAP n serve the frame $n + 1$ (as in Figure 9.8), identified as the maximum time relevance of DL-MAP and UL-MAP in Reference [1].
- UL-MAP n serves the frame n, identified as the minimum time relevance of DL-MAP and UL-MAP in Reference [1].

These two timings can be used for both the TDD and FDD variants of operation.

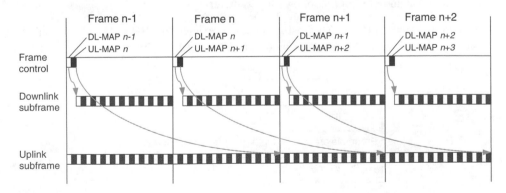

Figure 9.8 DL-MAP and UL-MAP indicate the use of downlink and uplink subframes (the FDD mode). (From IEEE Std 802.16-2004 [1]. Copyright IEEE 2004, IEEE. All rights reserved.)

Management message type of DL-MAP (= 2); (8 bits)	DCD count (8 bits)	Base Station ID (48 bits)	CID=i_1 DIUC=j_1 Start Time=t_1	CID=i_2 DIUC=j_2 Start Time=t_2	CID=i_3 DIUC=j_3 Start Time=t_3	...

DL-MAP IE$_1$ DL-MAP IE$_2$ DL-MAP IE$_3$

Figure 9.9   DL-MAP MAC management message general form for OFDM PHY. Each DL-MAP IE indicates the start time of a downlink burst and the burst profile (channel details including physical attributes) of this burst.

9.4.1 DL-MAP Message

The DL-MAP is a MAC management message that defines burst start time and profiles on the downlink. Each burst start time is indicated by a DL-MAP_IE (DL-MAP Information Elements). The DL-AMP_IE format is PHY layer-dependent. The BSs generate OFDM PHY DL-MAP messages in the format shown in Figure 9.9, including all of the following parameters:

- MAC management message type (= 2 for DL-MAP).
- PHY synchronisation. The PHY synchronisation field is dependent on the PHY specification used. This field is empty (zero bytes long) for the OFDM PHY Layer.
- DCD count. The value of the Configuration Change Count (CCC field) of the DCD, which describes the downlink burst profiles concerned by this map.
- Base Station ID. The Base Station ID is a 48-bit long field identifying the BS. The Base Station ID is programmable: the most significant 24 bits are used as the operator ID. This is a network management hook that can be combined with the Downlink Channel ID of the DCD message for handling edge-of-sector and edge-of-cell situations. Evidently, this is not the MAC address of the BS.

The remaining part of a DL-MAP is the encoding of the DL-MAP IEs that are PHY-specification dependent. The DL-MAP IE of the OFDM PHY Layer has the format shown in Figure 9.10 and includes all of the following parameters:

- Connection IDentifier (CID). This realises the assignment of the IE to a broadcast, multicast or unicast address. If the broadcast or multicast CID is used then it is possible to concatenate

CID	16 bits
DIUC	4 bits
Preamble present	1 bit (0=not present, 1=present)
Start Time	11 bits

Figure 9.10   DL-MAP IE fields for the OFDM (WiMAX) PHY Layer.

unicast MAC PDUs (with different CIDs) into a single downlink burst. During a broadcast
or multicast downlink burst, it is the responsibility of the BS to ensure that any MAC PDUs
sent to an H-FDD SS do not overlap any uplink allocations for that SS. An H-FDD SS for
which a DL-MAP_IE and UL-MAP_IE overlap in time uses the uplink allocation and discards downlink traffic during the overlapping period.

- DIUC. The 4-bit DIUC defines the burst type associated with that burst time interval. Burst
  profile descriptions are part of the DCD message for each DIUC used in the DL-MAP
  except those associated with Gap, End of Map and Extended (see DIUC in Section 9.5
  below).
- Preamble present. If set, the indicated burst will start with the short preamble (see above for
  preambles). In the downlink, a short preamble can be optionally inserted at the beginning
  of a downlink burst in addition to the long preamble that exists by default at the beginning
  of the frame.
- Start Time. This indicates the start time, in units of OFDM symbol duration, relative to the
  start of the first symbol of the PHY PDU (including the preamble) where the DL-MAP message is transmitted. The time instant indicated by the Start Time value is the transmission
  times of the first symbol of the burst including the preamble (if present). The end of the
  last allocated burst is indicated by allocating an End of Map burst (DIUC = 14) with zero
  duration (see the DIUC part in Section 9.5.5).

If the length of the DL-MAP message is a nonintegral number of bytes, the LEN field in
the MAC header is rounded up to the next integral number of bytes. The message is padded
to match this length, but the SS disregards the pad bits.

9.4.2 UL-MAP Message

The UL-MAP message allocates access to the uplink channel. The general format of the UL-
MAP message is almost identical to DL-MAP and is shown in Figure 9.11. There is only one
new field: the Allocation Start Time, which is the start time of the uplink allocation. The unit
of the Allocation Start Time is the PS starting from the beginning of the downlink frame in
which the UL-MAP message is placed. For the OFDM PHY, the minimum value specified for
this parameter is defined as the point in the frame 1 ms after the last symbol of the UL-MAP.
The Start Time field is in units of OFDM symbol duration (as for DL-MAP_IEs).

Management message type of UL-MAP (= 3); (8 bits)	UCD count (8 bits)	Base Station ID (48 bits)	Allocation Start time (32 bits)	CID=i_1 UIUC=j_1 Start Time=t_1 Duration= D_1	CID=i_2 UIUC=j_2 Start Time=t_2 Duration= D_2	...
				UL-MAP IE$_1$	UL-MAP IE$_2$	

Figure 9.11   UL-MAP MAC management message general form. For the sake of simplicity, not all the
fields are shown in this figure. Each UL-MAP IE indicates the start time of an uplink burst and the burst
profile (channel details including physical attributes) of this burst.

UL-MAP IE has some new elements with regard to DL-MAP:

- Duration indicates the duration, in units of OFDM symbols, of the allocation. The duration is inclusive of the preamble, the midambles and the postamble, contained in the allocation.
- Subchannel Index corresponds to the frequency offset indices of the subcarriers of the allocated subchannel (see Chapter 5 for OFDM subchannellisation).
- Midamble Repetition Interval indicates whether there is a midamble and, if there is one, the midamble repetition interval in OFDM symbols (8, 16 or 32 data symbols). When the last section of the symbol after the last midamble is longer than half the midamble repetition interval, a postamble must be added at the end of the allocation.

When specifically addressed to allocate a bandwidth grant to an SS, the CID is the Basic CID of the SS. A detailed example of a UL-MAP message with numerical values is given in Chapter 10. Contentions slots at the beginning of an uplink subframe are included in this example.

Within a frame, the switch from non-AAS to AAS-enabled traffic is marked by using UIUC = 15 with the AAS_IE to indicate that the subsequent allocation until the end of the frame is for AAS traffic [1]. Stations not supporting the AAS functionality ignore the portion of the frame marked for AAS traffic.

9.4.3 OFDMA PHY UL-MAP and DL-MAP Messages

The DL-MAP IEs and UL-MAP IE are PHY-specification dependent. The OFDMA DL-MAP IE defines a two-dimensional allocation pattern instead of one for OFDM DL-MAP IE. The OFDMA DL-MAP IE parameters are shown in Table 9.2.

The OFDMA UL-MAP IE has almost the same parameters as the OFDMA DL-MAP IE. A parameter proper to OFDMA UL-MAP IE is the Ranging Method parameter, which indicates one of two possible ranging bursts:

- Initial Ranging/Handover Ranging;
- BW Request/Periodic Ranging.

Table 9.2 OFDMA DL-MAP IE main parameters

DIUC	Used for the burst
OFDMA symbol offset	The offset of the OFDMA symbol in which the burst starts, measured in OFDMA symbols from the beginning of the downlink frame in which the DL-MAP is transmitted
Subchannel offset	The lowest index OFDMA subchannel used for carrying the burst, starting from subchannel 0
Boosting	Indication of whether the subcarriers for this allocation are power boosted
Number of OFDMA triple symbols	The number of OFDMA symbols that are used (fully or partially) to carry the downlink PHY burst. The value of the field is given in multiples of three symbols
Number of subchannels	The number of subchannels with subsequent indexes used to carry the burst
Repetition coding indication	Indicates the repetition code used inside the allocated burst

9.5 Burst Profile Usage: DCD Message and the DIUC Indicator

In this section, the mechanism for burst profile transmission and the main parameters required for this operation are described. A burst is a contiguous portion of data stream using
the same PHY parameters. These parameters are modulation type, Forward Error Correction
(FEC) type, preamble length, etc. (see below), and are known as the burst profile. This type
of multiplexing is known as TDM (Time Division Multiplexing) in 802.16. Each burst profile
is indicated by its Downlink Interval Usage Code (DIUC) or Uplink Interval Usage Code
(UIUC) for the downlink and uplink respectively.

9.5.1 Burst Profile Selection Thresholds

The burst profile attributes were described in Chapter 6. The main parameter is the Modulation and Coding Scheme (MCS) used in the burst. In addition, and as seen in Chapter 6 and
Annex B, two CINR threshold numerical values are provided with each burst profile description: the DIUC mandatory exit threshold and the DIUC mandatory entry threshold. These
two thresholds are used for the selection of one of the defined (in the DCD message) burst
profiles, based on the received CINR. The procedure is illustrated in Figure 9.12 and stated
below:

- If CINR ≤ DIUC mandatory exit threshold
     → this DIUC can no longer be used
     → change to a more robust DIUC is required.
- If CINR ≥ DIUC minimum entry threshold
     → this DIUC can be used
     → change to a less robust DIUC.

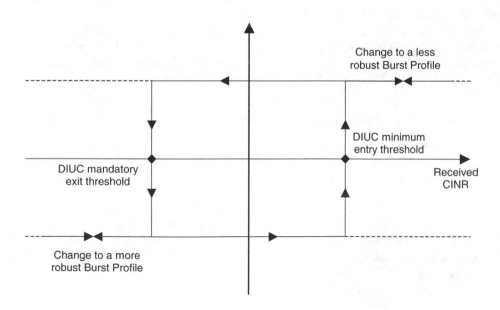

Figure 9.12   Use of thresholds for a given burst profile.

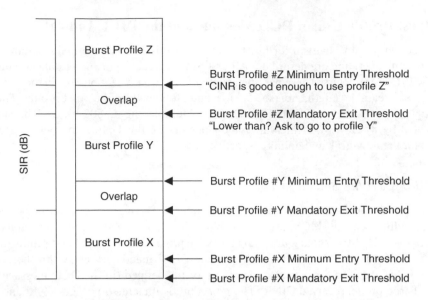

Figure 9.13 Illustration of burst profile threshold values for three neighbouring profiles. Burst profile X is the most robust. (From IEEE Std 802.16-2004 [1]. Copyright IEEE 2004, IEEE. All rights reserved.)

The possible thresholds values for neighbouring burst profiles are shown in Figure 9.13. The downlink burst profile is determined by the BS according to the quality of the signal that is received by each SS. While operating on a given burst profile, the SS monitors its received CINR and compares the average value with the allowed range of operation. This region is bounded by the two threshold levels of the burst profile used. These two threshold levels are defined along with the burst profile in the DCD message.

If the received CINR goes outside the allowed operating region (from the top or from the bottom), the SS must request a change to a new burst profile, according to the principle given above in this subsection and using one of two proposed methods (described in Chapter 11). This depends on whether this SS has been given an uplink data grant allocation or not.

9.5.2 DCD (Downlink Channel Descriptor) Message

The DCD (Downlink Channel Descriptor) message is a broadcasted MAC management message transmitted by the BS at a periodic time interval in order to provide the burst profiles (physical parameter sets) that can be used by a downlink physical channel during a burst, in addition to other useful downlink parameters. The general format of a DCD messages is shown in Figure 9.14. The contents of the DCD message fields are now given.

The Downlink Channel ID (ChID, not to be confused with CID) is the identifier of the downlink channel to which this message refers. This identifier is arbitrarily chosen by the BS and is unique only within the MAC domain. This acts as a local identifier for transactions such as ranging.

The Configuration Change Count (CCC) is an 8 bits field, incremented by one (modulo 256) by the BS whenever any of the values of the DCD message changes. If the value of this

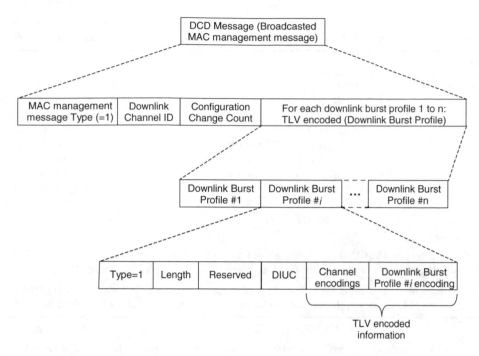

Figure 9.14   General format of the DCD (Downlink Channel Descriptor) message

count in a subsequent DCD remains the same, the SS can decide, after reading the CCC field, that the remaining fields have not changed. The SS may then disregard the remainder of the DCD message.

The Downlink_Burst_Profile fields describe each of the burst profiles that can be used in the downlink until the next DCD message. The Downlink_Burst_Profile field contents are defined separately for each PHY specification (Single Carrier, OFDM, OFDMA, etc.). Table 9.3 defines the format of the Downlink_Burst_Profile encoding, which is used in the DCD message of the OFDM (WiMAX) interface. This is the format shown in Figure 9.14. The Downlink_Burst_Profile includes a TLV encoding (see Section 8.6) that defines and associates with a particular Downlink Interval Usage Code (DIUC) value, the PHY characteristics (burst profile) used with that DIUC. Within each downlink burst profile is a list of PHY attributes (burst profile encodings) for that burst profile, encoded as TLV values.

Table 9.3   Format of the downlink burst profile for the OFDM (WiMAX) profile

Field	Size
Type = 1	8 bits
Length	8 bits
Reserved	4 bits
DIUC	4 bits
TLV encoded information	Variable

Table 9.4 Some parameters of an OFDM PHY burst profile: FEC and modulation possible values, the corresponding mandatory CINR thresholds values (see Chapter 6 for burst profile parameters)

Name	Value	
FEC code type	0 = BPSK (CC) 1/2	11 = 64-QAM (BTC) 2/3
	1 = QPSK (RS + CC/CC) 1/2	12 = 64-QAM (BTC) 5/6
	2 = QPSK (RS + CC/CC) 3/4	13 = QPSK (CTC) 1/2
	3 = 16-QAM (RS + CC/CC) 1/2	14 = QPSK (CTC) 2/3
	4 = 16-QAM (RS + CC/CC) 3/4	15 = QPSK (CTC) 3/4
	5 = 64-QAM (RS + CC/CC) 2/3	16 = 16-QAM (CTC) 1/2
	6 = 64-QAM (RS + CC/CC) 3/4	17 = 16-QAM (CTC) 3/4
	7 = QPSK (BTC) 1/2	18 = 64-QAM (CTC) 2/3
	8 = QPSK (BTC) 3/4 or 2/3	19 = 64-QAM (CTC) 3/4
	9 = 16-QAM (BTC) 3/5	20–255 = Reserved
	10 = 16-QAM (BTC) 4/5	
DIUC mandatory exit threshold	The CINR at or below where this DIUC can no longer be used and where a change to a more robust DIUC is required. Expressed in 0.25 dB units. See Section 9.5.1.	
DIUC minimum entry threshold	The minimum CINR required to start using this DIUC when changing from a more robust DIUC is required. Expressed in 0.25 dB units. See Section 9.5.1.	

The TLV encoded information of a burst profile is made of two fields (see Figure 9.14):

- Channel encodings. The main parameters are the BS EIRP (Effective Isotropic Radiated Power), the TTG (Tx/Rx Transition Gap) and the RTG (Receive/transmit Transition Gap), the BS ID, the frame duration, etc. See the example in Section 9.5.4.
- The downlink burst profile number i encoding. This indicates the type of modulation and channel coding and the received signal thresholds required for using this profile, which will be indicated by DIUC #i. As already mentioned, four modulations can be used (BPSK, QPSK, 16-QAM and 64-QAM) with four possible coding types (RS, CC, BTC and CTC). Table 9.4 shows possible values of a burst profile as defined by the 802.16 standard for the OFDM PHYsical interface.

9.5.3 Transmission of the DCD Message

For all 802.16 PHYsical Layers, the maximum value of the time between the transmissions of two consecutive DCD messages (or two consecutive UCD messages) is 10 s. For OFDM and OFDMA PHY, the frame duration is between 2 ms and 20 ms (although all the possible values are not necessarily mandatory for WiMAX profiles).

It can then be concluded that a non-DCD-including frame is transmitted more frequently than a DCD message. As already mentioned, each frame may contain many bursts, each burst having a unique DIUC. Therefore, each DCD message will concern a large number of frames and an even larger number of bursts (see Figure 9.15).

9.5.4 An Example of the DCD Message

An example will now be given of a DCD MAC management message. In this message, two burst profiles are defined and associated with DIUC 0101 (hexadecimal: 5) and DIUC 1010 (hexadecimal: A).

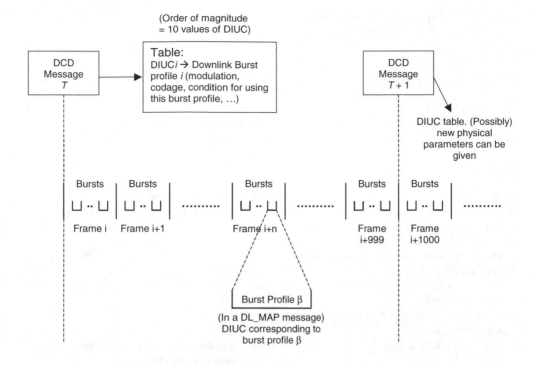

Figure 9.15   Illustration of the DCD message transmission period and DIUC use. The value of 1000 frames between two DCD messages is an order of magnitude.

The global parts of this message are shown in Table 9.5. The full details of this message, including fields lengths, are given in Annex B. For this message the OFDM PHYsical interface specifications are considered. 802.16e added new parameters to the DCD message (mainly for handover process); they are not taken into account in this example.

9.5.5 DIUC Values

The DIUC (Downlink Interval Usage Code) is then the indicator of a burst profile, i.e. the PHY characteristics (modulation, encoding, burst profile use condition, etc.) of a downlink burst. The DIUC is a 4-bit field. The value of DIUC is PHY layer-dependent. Table 9.6 shows the values defined for the OFDM (WiMAX) PHY Layer. Only 11 values are used for burst profile selection. The correspondence between the 20 modulation and coding scheme possibilities shown in Table 9.5 and the DIUC value (a maximal number of 11 burst profile value indicators) is the choice of the BS.

Each interval or, more specifically, burst (downlink or uplink) will have its burst profile and start time described by a DL-MAP IE (Downlink MAP Message Information Element) or a UL-MAP IE. The DL-MAP and the UL-MAP are MAC management messages that describe the use of the time frame by different SSs (see above). The burst profile is referenced by the DIUC value. Only bursts whose profile is explicitly known, which is the case for some control bursts (example: FCH burst, see above), do not have a burst profile DIUC value.

Table 9.5 Example of a DCD message containing two burst profile descriptions (OFDM PHY, 802.16e modifications not included). The full details are given in Annex B

Field contents	Description
MAC management message Type (=1)	Identification of the MAC management message = DCD
Downlink Channel ID	
Configuration Change Count	Indication of possible DCD change
Type = 1	Start of downlink burst profile 0101 (OFDM PHY Layer format)
Length	
Reserved	
DIUC = 0101	DIUC value indicating this burst profile
TLV of downlink burst profile 0101, indicating the length of this object	Start of downlink burst profile 0101 channel encodings (OFDM PHY Layer format)
TLV of the BS transmitted power	
TLV of the TTG (transmit burst/receive burst transition gap)	
TLV of Base Station ID	
TLV of Frame Duration	
Other TLVs of downlink burst profile 0101 channel encodings	
TLV of downlink frequency	Start of downlink burst profile 0101 burst profile encodings (OFDM PHY Layer format)
TLV of coding and modulation scheme (called FEC code)	
TLV of DIUC selection thresholds	
Other TLV of downlink burst profile 0101 burst profile encodings	
Type = 1	Start of downlink burst profile 1010 (OFDM PHY Layer format)
Same fields as for downlink burst profile 0101 (with possible different values)	

Table 9.6 The possible values of DIUC (coded on 4 bits) for OFDM PHYsical Layer. Only 11 values are used for burst profile selection

DIUC	Usage
0	STC zone
1–11	Burst profiles
12	*Reserved*
13	Gap
14	End of Map
15	Extended DIUC

Table 9.7   Extended DIUC possible uses for the OFDM PHYsical Layer

Extended DIUC value	Possibility usage
0 × 00	Issued by the BS to request a channel measurement report. The Channel_Measurement_IE is followed by the End of Map IE
0 × 01	Indicates that the subsequent bursts utilise a preamble which is cyclically delayed in time by M samples (Physical Modifier IE)
0 × 02	Switch from non-AAS to AAS-enabled traffic. AAS, Adaptive Antenna System (see Chapter 12)
0 × 03	Specify one of a set of parallel downlink bursts for transmission (concurrent transmission IE format)
0 × 04	Indicate that the subsequent allocations, until the end of the frame, are STC encoded
0 × 05	Indicate that subsequent allocations use downlink subchannelisation (for a downlink subchannelisation-enabled BS)
0 × 06 → 0 × 0F	These extended DIUC values are called the Dummy IE. Left for future specifications

The DIUC possible values, other than burst profiles, shown in Table 9.6 are the following:

- STC (Space-Time Coding) is a transmission technique used to decrease multipath effects. The modulation used is QPSK.
- Gap is a period of time between the downlink burst and the subsequent uplink burst or between the uplink burst and the subsequent downlink burst. This gap allows time for the BS or the SS to switch from transmits to the receive mode and inversely.
- An End of Map IE terminates all allocations in an IE list. The end of the last allocated burst is indicated by allocating an End of Map burst.
- Extended DIUC. A DIUC value of 15 indicates that the IE carries special information. An Extended DIUC field, on 4 bits, is then present, showing the extended DIUC signification (see Table 9.7 for the OFDM PHYsical Layer).

An SS will ignore an IE with an extended DIUC value for which the station has no knowledge (e.g. an SS that has no support for STC). In the case of a known extended DIUC value, but with a length field longer than expected, the SS processes the information up to the known length and ignores the remainder of the IE.

Table 9.8 shows the values defined for the OFDMA (WiMAX) PHY Layer. A disadvantage of an OFDM transmission is that it can have a high PAPR (Peak to Average Power

Table 9.8   The possible values of DIUC for the OFDMA PHYsical Layer

DIUC	Usage
0–12	Burst profiles
13	Gap/PAPR
14	Extended-2 DIUC
15	Extended DIUC

Ratio). The PAPR is the peak value of transmitted subcarriers to the average transmitted signal. A high PAPR represents a hard constraint for some devices (such as amplifiers). DIUC = 13 may be used for the allocation of subchannels for PAPR reduction schemes. The subcarriers within these subchannels may be used by all SSs to reduce the PAPR of their transmissions. The SS will ignore the received signal (subcarriers) in the GAP/PAPR reduction region.

9.5.6 UCD (Uplink Channel Descriptor) Message and UIUC Indicator

The UCD (Uplink Channel Descriptor) message is a broadcasted MAC management message transmitted by the BS at a periodic time interval in order to provide the burst profile (physical parameter sets) description that can be used by an uplink physical channel in addition to other useful uplink parameters. Its functioning is very similar to the DCD so will not be described in as much detail.

A UCD message must be transmitted by the BS at a periodic interval in order to define the characteristics of an uplink physical channel. The maximum allowed value for this period is 10 s (as for DCD). The UCD message of OFDM PHY includes the following parameters:

- Configuration Change Count. This is the same as for DCD.
- Ranging Backoff Start and Ranging Backoff End (8 bits each). These are initial backoff and final (or maximum) backoff window sizes for initial ranging contention (see Chapter 11), expressed as a power of 2. Values of these exponents are in the range 0–15.
- Request Backoff Start and Request Backoff End (8 bits each). These are initial backoff and final (or maximum) backoff window sizes for contention BW (bandwidth) requests (see Chapter 10), expressed as a power of 2. Values of these exponents are in the range 0–15.
- For each uplink burst profile defined in this UCD message, Uplink_Burst_Profile, which is a compound TLV encoding that defines and associates with a particular UIUC, the PHY characteristics that must be used with that UIUC. The TLV encoded values of a burst profile are globally similar to the ones of the downlink burst profiles in the DCD message. The following ones are burst profile parameters specific to UCD:
- Contention-based reservation timeout. This is the number of UL-MAPs received before a contention-based reservation is attempted again for the same connection.
- Bandwidth request opportunity size. This is the size (in units of PS) of the PHY payload that an SS may use to format and transmit a bandwidth request message in a contention request opportunity. The value includes all PHY overhead as well as allowance for the MAC data the message may hold.
- Ranging request opportunity size. This is the size (in units of PS) of the PHY bursts that an SS may use to transmit a Ranging Request message in a contention ranging request opportunity (see Chapter 11). The value includes all PHY overheads and (in addition to the bandwidth request opportunity size content) the maximum SS/BS round trip propagation delay.
- Subchannelisation REQ Region-Full Parameters. This is the number of subchannels used by each transmit opportunity when REQ Region-Full is allocated in a subchannelisation region. Possible values are between 1 and 16 subchannels (see Section 10.4).
- Subchannelisation focused contention codes. This is the number of contention codes (C_{SE}) that can be used to request a subchannelised allocation. The default value is 0 (no

Table 9.9 The possible values of the UIUC (coded on 4 bits) for OFDM PHY

UIUC	Usage
0	*Reserved*
1	Initial ranging
2	REQ (Request) region full
3	REQ (Request) region focused
4	Focused contention IE
5–12	Burst profiles
13	Subchannelisation network entry
14	End of Map
15	Extended UIUC

subchannelised focused contention). Allowed values are between 0 and 8. Focused contention is described in Section 10.4.

As for the DIUC and the DCD, the UIUC (Uplink Interval Usage Code) is defined as an indicator of one of the uplink burst profiles described in the UCD. The UIUC is a 4-bit field corresponding to 16 possible values. The value of UIUC is PHY layer-dependent. Table 9.9 shows the UIUC values defined for the OFDM (WiMAX) PHY layer. Only eight values are used for burst profile selection. The UL-MAP IE for allocation of bandwidth in response to a subchannelised network entry signal (see Chapter 10), in the subchannelised section of the UL-MAP, is identified by UIUC = 13. An SS responding to a bandwidth allocation using the subchannelised network entry IE starts its burst with a short preamble and uses only the most robust mandatory burst profile in that burst.

There are 20 available modulation and coding schemes for uplink burst profiles. The most robust is BPSK with a channel coding rate of 1/2 and the less robust being 64-QAM with a coding rate of 5/6 (both OFDM and OFDMA layers). The correspondence between these 20 available modulation and coding schemes for uplink burst profiles and the UIUC value is the choice of the BS. Only eight UIUC values can be used as indicators of uplink burst profiles (equivalently, only eight uplink burst profiles may be defined in an UCD).

Many of the UIUC values shown in Table 9.9 will be used in the following chapters. The initial ranging process is described in Chapter 11. Uplink bandwidth request procedures (concerning UIUC values 2 to 4) are described in Chapter 10. The value 13 of UIUC corresponds to the subchannelised network entry IE, used in the procedure of subchannelisation network entry. Extended DIUC allows additional functions. For example, when a power change for the SS is needed, UIUC = 15 is used with an extended UIUC set to 0×00 and with an 8-bit power control value. This power control value is an 8-bit signed integer expressing the change in power level (in 0.25 dB units) that the SS must apply to correct its current transmission power.

For OFDMA PHY, the sounding zone is a region of one or more OFDMA symbol intervals in the uplink frame that is used by the SS to transmit sounding signals to enable the BS to determine rapidly the channel response between the BS and the SS. The BS may command an SS to transmit a sounding signal at one or more OFDMA symbols within the sounding zone by transmitting the UL-MAP message UL_Sounding_Command_IE( ) to provide detailed sounding instructions to the SS. In order to enable uplink sounding, in UL-MAP, a

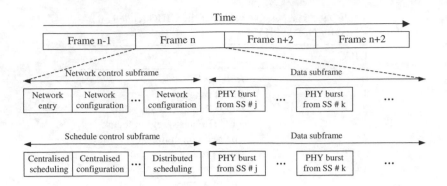

Figure 9.16 Mesh frame global structure. According to the standard, Mesh networks can only use the TDD mode

BS transmits UIUC = 13 with the PAPR_Reduction_Safety_and_Sounding_Zone_Allocation_IE( ) to indicate the allocation of an uplink sounding zone within the frame.

9.6 Mesh Frame

The PMP topology supports both TDD and FDD duplexing modes, while Mesh topology supports only the TDD duplexing mode. In the case of a Mesh network, on the opposite side of the basic PMP mode, there can be no separate downlink and uplink subframes since all stations have the same hierarchy. An (optional) Mesh frame structure is defined in the 802.16 standard to facilitate Mesh networks. Figure 9.16 shows the global structure of this Mesh (TDD) frame. The contents of this Mesh frame are now described.

A Mesh frame consists of a control and a data subframe. This frame uses information contained in the MAC management message MSH-NCFG (Mesh Network Configuration) and, specifically, the Network Descriptor IE.

The control subframe serves two basic functions. The first function is defined as network control and realises the creation and maintenance of cohesion between the different systems. It is described in Section 9.6.1 below. The other function is defined as schedule control and realises the coordinated scheduling of data transfers between systems. It is described in Section 9.6.2. Frames with a network control subframe occur periodically, as indicated in the Network Descriptor, included in this subframe and detailed below. All other frames have a schedule control subframe. The length of the control subframe is fixed and of length MSH-CTRL-LEN × 7 OFDM symbols, where MSH-CTRL-LEN is a parameter indicated in the Network Descriptor IE of MSH-NCFG.

9.6.1 Network Control Subframe

The Network Control subframe is made of two parts and is shown in Figure 9.17. The MAC PDUs of these two parts, the network entry and the network configuration, contain two Mesh messages: MSH-NENT and MSH-NCFG:

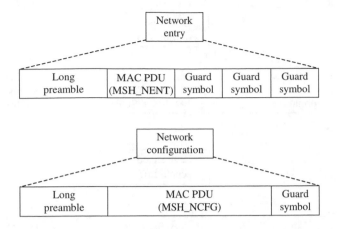

Figure 9.17   The two parts of the Network Control subframe of the Mesh subframe. The network configuration contains the Network Descriptor

- MSH-NENT (Mesh Network Entry) is a basic MAC management message that provides the means for a new node to gain synchronisation and initial network entry into a Mesh network.
- MSH-NCFG (Mesh Network Configuration) is a broadcasted MAC management message that provides a basic level of communication between nodes in different nearby networks, whether from the same or different equipment vendors or wireless operators. Among others, the Network Descriptor is an embedded data of the MSH-NCFG message. The Network Descriptor contains many channel parameters (modulation and coding schemes, threshold values, etc.), which makes it similar to the UCD and DCD.

9.6.2 Schedule Control Subframe

The Schedule Control subframe is made of three parts and is shown in Figure 9.18. The MAC PDUs of these three parts, the centralised configuration, the centralised scheduling and the distributed scheduling contain three Mesh messages: MSH-CSCF, MSH-CSCH and MSH-DSCH:

- MSH-CSCF (Mesh Centralised Schedule Configuration) and MSH-CSCH (Mesh Centralised Schedule) are broadcasted MAC management messages that are broadcasted in the Mesh mode when using centralised scheduling. The Mesh BS broadcasts these messages to all its neighbours and all nodes forward (rebroadcast) them.
  The Mesh BS may create a MSH-CSCH message and broadcast it to all its neighbours to grant bandwidth to a given node, and then all the nodes with a hop count lower than a given threshold forward the MSH-CSCH message to their neighbours that have a higher hop count. On the other hand, nodes can use MSH-CSCH messages to request bandwidths from the Mesh BS. Each node reports the individual traffic demand requests of each 'child' node in its subtree to the Mesh BS.

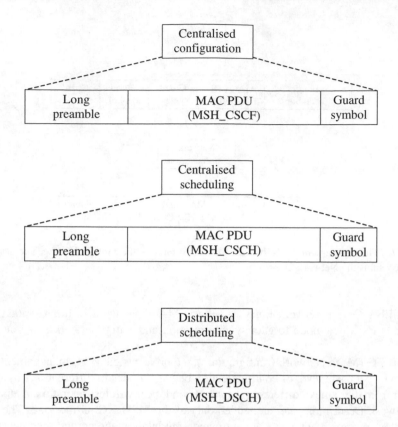

Figure 9.18   The three parts of the Schedule Control subframe of the Mesh subframe

- MSH-DSCH (Mesh Distributed Schedule) is a broadcasted MAC management message that is transmitted in the Mesh mode when using distributed scheduling. In coordinated distributed scheduling, all the nodes transmit a MSH-DSCH at regular intervals to inform all the neighbours of the schedule of the transmitting station. The coordination protocol is provided in the standard. Further, the MSH-DSCH messages are used to convey information about free resources, indicating where the neighbours can issue grants.

10

Uplink Bandwidth Allocation and Request Mechanisms

10.1 Downlink and Uplink Allocation of Bandwidth

Downlink and uplink bandwidth allocations are completely different. The 802.16 standard has a MAC centralised architecture where the BS scheduler controls all the system parameters, including the radio interface. It is the role of this BS scheduler to determine the uplink and downlink accesses. The uplink and downlink subframe details were given in Chapter 9.

The downlink allocation of bandwidth is a process accomplished by the BS according to different parameters that are determinant in the bandwidth allocation. Taking into consideration the QoS class for the connection and the quantity of traffic required, the BS scheduler supervises the link and determines which SS will have downlink burst(s) and the appropriate burst profile. In this chapter, the uplink access mechanisms of WiMAX/802.16 are described. Chapter 11 describes scheduling and QoS.

In the uplink of each BS zone or, equivalently, WiMAX cell, the SSs must follow a transmission protocol that controls contention between them and enables the transmission services to be tailored to the delay and bandwidth requirements of each user application. This is accomplished while taking into account five classes of uplink service levels, corresponding to the five QoS classes that uplink transmissions may have.

Uplink access and bandwidth allocation are realised using one of the four following methods:

- unsolicited bandwidth grants;
- piggyback bandwidth request;
- unicast polling, sometimes simply referred to as polling;
- contention-based procedures, including broadcast or multicast polling, where contention-based bandwidth request procedures have variants depending of the PHYsical Layer used: OFDM or OFDMA (see below).

The standard states that these mechanisms are defined to allow vendors to optimise system performance by using different combinations of these bandwidth allocation techniques while maintaining consistent interoperability. The standard proposes, as an example, the use of contention instead of individual polling for SSs that have been inactive for a long period of time. Next, the realisation of these methods is described, but first two possible differentiations of an uplink grant-request are introduced.

10.2 Types of Uplink Access Grant-request

The BS decides transmissions in the uplink and the downlink. For uplink access, a grant is defined as the right for an SS to transmit during a certain duration. Requests for bandwidth must be made in terms of the number of bytes needed to carry the MAC header and payload, but not the PHY overhead. For an SS, bandwidth requests reference individual connections while each bandwidth grant is addressed to the SS's Basic CID, not to individual CIDs. It is then up to the SS to use the attributed bandwidth for any of its CIDs. Since it is nondeterministic which request is being honoured, when the SS receives a shorter transmission opportunity than expected due to a scheduler decision, the request message loss or some other possible reason, no explicit reason is given.

Grants are then given by the BS after receipt of a request from an SS. Two possible differentiations can be made for this request. These differentiations are now described.

10.2.1 Incremental and Aggregate Bandwidth Request

A grant-request (by an SS) may be incremental or aggregate:

- When the BS receives an incremental bandwidth request, it adds the quantity of bandwidth requested to its current perception of the bandwidth needs of the connection.
- When the BS receives an aggregate bandwidth request, it replaces its perception of the bandwidth needs of the connection with the quantity of bandwidth requested.

The self-correcting nature of the request-grant protocol requires that the SSs should periodically use aggregate Bandwidth Requests. The standard states that this period may be a function of the QoS of a service and of the link quality, but do not give a precise value for it.

The grant-request may be sent in two possible MAC frame types that are described in the following subsection. Only the first one (the standalone bandwidth request) can be aggregate or incremental.

10.2.2 Standalone and Piggyback Bandwidth Request

The two MAC frame types of the 802.16 standard, already defined in Section 8.2, can be used by an SS to request bandwidth allocation from the BS. Specifically, Section 8.2.3 details MAC headers and gives two types of request that are now described.

The standalone bandwidth request is transmitted in a dedicated MAC frame having a Header format without payload Type I, indicated by the first bit of the frame, the Header Type bit, being equal to 1. A Type field in the bandwidth request header indicates whether the request is incremental or aggregate (see Table 8.3). In the bandwidth request header, a 19-bit long bandwidth request field, the Bandwidth Request field, indicates the number of bytes of the uplink bandwidth requested by the SS for a given CID, also given in this header.

The standalone bandwidth request is included in the two main grant-request methods: unicast polling and contention-based polling.

For any uplink allocation, the SS may optionally decide to use the allocation for:

- data;
- requests;
- requests piggybacked in data.

The piggyback bandwidth request uses the grant management subheader which is transmitted in a generic MAC frame (then having a generic MAC header). This is indicated by the first bit of the frame, the Header Type bit, being equal to 0. This can avoid the SS transmitting a complete (bandwidth request MAC header) MPDU with the overhead of a MAC header only to request bandwidth. The grant management subheader, in a generic MAC frame, is used by the SS to transmit bandwidth management needs to the BS. The Type bit in the generic MAC frame header (see Table 8.2) indicates the possible presence of a grant management subheader.

The piggyback bandwidth request (grant management subheader) is a lightweight way to attach a request uplink bandwidth without having to create and transmit a complete MPDU with the overhead of MAC headers and CRCs. The grant management subheader is two bytes long. It is used by the SS to transmit bandwidth management needs to the BS in a generic MAC header frame in addition to other possible data transmitted in the same MAC frame. Depending on the class of QoS of the connection, three types of grant management subheader are defined and used:

1. Grant management subheader (see Figure 10.1). This is the case for QoS class = UGS. The UGS (Unsolicited Grant Services) is a QoS class designed to support real-time service flows, where the SS has a regular uplink access (for more details on the UGS class, see Chapter 11). For this class of QoS, the PM (Poll-Me) bit in the grant management subheader can be used by the SS to indicate to the BS that it needs to be polled in order to request bandwidth for non-UGS connections (see below).

    The grant management subheader contains only two useful bits, the SI (Slip Indicator), used by the SS to indicate a slip of uplink grants relative to the uplink queue depth, and the PM (Poll-Me) bit, used by the SS to request a bandwidth poll, probably for a needed additional uplink bandwidth with regard to the regular access this UGS SS has. The Slip

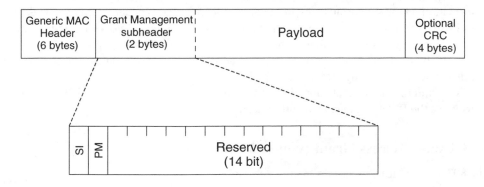

Figure 10.1   Grant management subheader for the QoS class = UGS (Unsolicited Grant Services)

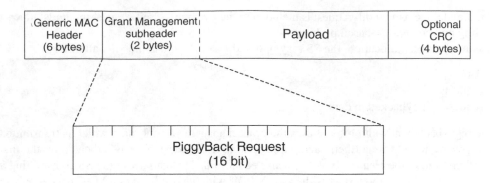

Figure 10.2   Grant management subheader for the QoS class ≠ UGS (Unsolicited Grant Services). The piggyback request field is the number of bytes of the uplink bandwidth requested by the SS

Indicator bit use by the SS is the following: the BS may provide for long-term compensation for possible bad conditions, such as lost maps or clock rate mismatches, by issuing additional grants. The SI flag allows prevention of the provision of too much bandwidth by the BS. The SS sets this flag once it detects that the service flow has exceeded its transmit queue depth. Once the SS detects that the service flow transmit queue is back within limits, it clears the SI flag. No precise values for these limits are given in the standard.

2. Piggyback grant-request subheader (see Figure 10.2). This is the case for the QoS classes ≠ UGS. Since the piggybacked bandwidth request subheader does not have a Type field, it will always be incremental. The piggyback request field is the number of bytes of the uplink bandwidth requested by the SS. The bandwidth request is for the CID indicated in the MAC frame header (see Section 8.2).

3. Extended piggyback request. This is defined by the 16e amendment along with and for the (newly defined) ertPS class. The number of bytes of the uplink bandwidth (piggyback) requested by the SS is on 11 bits (instead of 16). This request is incremental. In the case of the ertPS class, if the MSB (Frame Latency Indicator, or FLI) of the grant management subheader is 1, the BS changes its polling size into the size specified in the LSBs of this field (Frame Latency (FL) field).

The standard (16e) states that FL and FLI fields may be used to provide the BS with information on the synchronisation of the SS application that is generating periodic data for UGS/extended rtPS service flows. The SS may use these fields to detect whether latency experienced by this service flow at the SS exceeds a certain limit, e.g. a single frame duration. If the FL indicates inordinate latency, the BS may shift scheduled grants earlier for this service flow (taking into account the FL).

The standard states that capability of the piggyback request is optional. This probably includes the PM grant management subheader.

10.3 Uplink Access Grant-request Mechanisms

The 802.16 standard defines two main grant-request methods:

• unicast polling (or polling);
• contention-based polling.

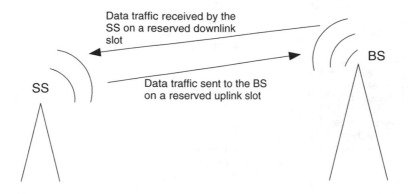

Figure 10.3   Unsolicited bandwidth grants in the uplink

By extension, the UGS class of QoS has unsolicited bandwidth grants, sometimes considered as an (implicit) grant-request mechanism although it is based on reserved slots dedicated for the concerned UGS class SSs. These grant-request mechanisms will now be described, starting with the simplest one, unsolicited bandwidth grants.

10.3.1 Unsolicited Bandwidth Grants

The unsolicited bandwidth grants technique consists of dedicated slots reserved for UGS class SSs. This type of bandwidth requests is useful for applications requiring a fixed rate data stream. Figure 10.3 illustrates the unsolicited bandwidth grant mechanism in the uplink and the downlink. This type of access grant is used only by the UGS class of QoS.

10.3.2 Unicast Polling

Polling is the process by which the BS allocates bandwidth to the SSs for the purpose of making bandwidth requests. These allocations may be to an individual SS or to a group of SSs. The use of polling simplifies the access operation and guarantees that applications can receive service on a deterministic basis if it is required. This allocation technique is used when bandwidth resource demand is not relevant enough to have unsolicited bandwidth grants for all users; the BS can then directly assign the request amount to the SS(s) as needed.

When an SS is polled individually, it is a unicast polling. In the case of unicast polling, no explicit message is transmitted to poll the SS. Rather, the SS is allocated, in the UL-MAP, sufficient bandwidth to respond with a Bandwidth (BW) request. The standard indicates that for any individual uplink allocation, the SS may optionally decide to use the allocation for data, requests or requests piggybacked in data transmission. Taking into account its (possibly) different pending uplink transmission requests, the SS scheduler decides if a bandwidth request must be made, standalone or piggybacked with data (see Section 10.2). If the SS do not have data to transmit and then no need for bandwidth, the allocation is padded, eventually using a padding CID (see Table 7.1). Figure 10.4 represents the unicast polling mechanism.

The standard states that unicast polling would normally be done on a per-SS basis by allocating a Data Grant IE (or Data Grant Burst Type IE) directed at its Basic CID. A Data Grant

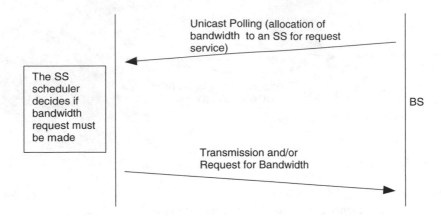

Figure 10.4 Illustration of the unicast polling mechanism. If the SS has no needs, the allocated slots are padded

IE (or Data Grant Burst Type IE) is a UL-MAP_IE with the UIUC indicating the burst profile of the uplink access duration allocated to an SS.

The SSs with currently active UGS connections may set the PM bit in the grant management subheader (see Figure 10.1) in the MAC packet of their UGS connection to indicate to the BS that they need to be polled to the request bandwidth for one or more non-UGS connection(s). To reduce the individual polling bandwidth requirements in the downlink, SSs with active UGS connections need to be polled individually only if this PM bit is set. Once the BS detects this request for polling, it applies the individual polling process.

10.3.3 Contention-based Group (Multicast or Broadcast) Polling

The available bandwidth may not be sufficient to individually poll all inactive SSs. Contention-based grant-request mechanisms are allocated a small part of each uplink frame (in the FDD mode) or subframe (in the TDD mode), known as the bandwidth requests contention slot (see Chapter 9, Figure 9.7). The size of this contention slot, known in the standard as the Request IE, is indicated by the BS (see Section 10.3.7). With this contention slot, an SS can access the network by asking the BS for an uplink slot. If the BS receives the demand (which means that there was no collision), it evaluates the SS request in the context of its service-level agreement, the radio network state and the scheduling algorithm, and possibly allocates a slot in which the SS can transmit data. Some SSs, such as those inactive for a long period of time and/or with low access priority, may then be polled in multicast groups. In some cases, a broadcast poll may also be made. Thus, multicast polling saves the bandwidth with regard to the scheme where all SSs are polled individually. In the case where this polling is made to a group of SSs, the allocated bandwidth is specifically for the purpose of making bandwidth requests.

Some CIDs are reserved for multicast groups and for broadcast messages (see Table 7.1). As for individual (unicast) polling, the poll is not an explicit message, but rather bandwidth allocated in the UL-MAP. The difference is that, rather than associating an allocated bandwidth with an SS's Basic CID, the allocation is to a multicast or broadcast CID. An example of a BS polling is provided in Section 10.3.7.

Group (multicast or broadcast) polling works as follows. When the poll is directed at a multicast CID or the broadcast CID, an SS belonging to the polled group may request a bandwidth during any request interval allocated to that CID in a UL-MAP. Figure 10.5 represents an illustration of the contention-based group polling mechanism. In order to reduce the likelihood of collision with multicast and broadcast polling, only SSs needing a bandwidth reply. These replying SSs apply a contention resolution algorithm, described in Section 10.3.5, to select the slot in which to transmit the initial bandwidth request. This mechanism allows a fair distribution of the bandwidth between different SSs without allocating a dedicated slot for each SS.

A replying SSs assumes that the transmission has been unsuccessful if it does not receive a grant after a given number of subsequent UL-MAP messages. This parameter, called the contention-based reservation timeout, is given in the UCD MAC management message (see Chapter 9 for a UCD message). If necessary, an SS transmits during the total time of all of its uplink grants using a given padding mechanism.

10.3.4 Management of Multicast Polling Groups

The BS may add an SS to a multicast polling group, identified by a multicast polling CID value, by sending the MCA-REQ (Multicast Polling Assignment Request) MAC management

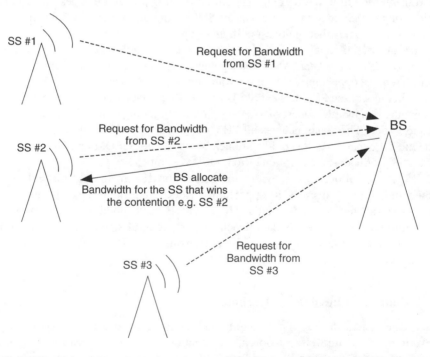

Figure 10.5 Illustration of contention-based group polling. The three SSs shown are group (multicast or broadcast) polled. They all have a bandwidth request. SS 2 wins the contention and then receives a bandwidth allocation

message with the Join command. On the other hand, the BS can remove an SS from a multicast polling group by sending the MCA-REQ MAC management message with the Leave command. Upon receiving the MCA-REQ message, the SS will respond by sending the MCA-RSP (Multicast Polling Assignment Response) MAC management message. Among the MCA-REQ MAC management message TLV parameters are the following: multicast CID (that the SS must join or leave) and assignment (leave or join). Multicast groups may have a periodic polling allocation after a number of frames indicated (and TLV coded) in the MCA-REQ message. This type of periodic polling (REQ Region Full or REQ Region Focused, see Section 10.4) is also among the MCA-REQ TLV parameters.

The MCA-RSP is sent by the SS in response to an MCA-REQ and contains mainly the confirmation code equal to zero if the request was successful and to non-zero in case of failure. These two messages use the primary management connection; i.e. they are sent on the SS's primary management CID (in the generic MAC header CID field).

10.3.5 Contention Resolution for Group Polling

10.3.5.1 Transmission Opportunity

A transmission opportunity in a contention-based procedure of 802.16 is defined as a contention space allocation provided in a UL-MAP for a group of SSs. In OFDM PHY, there are transmission opportunities dedicated to the transmission of bandwidth requests and others for the transmission of initial ranging. The initial ranging procedure is described in Chapter 11. This group of SSs may include either all SSs having an intention to join the cell or all registered SSs or some other multicast polling group.

The size of an individual transmission opportunity for each type of contention IE is indicated by the BS in the UCD MAC management message. This parameter, known as the Bandwidth request opportunity size or the Ranging request opportunity size (see Chapter 9), is the size in units of the PS (Physical Slot) of the PHY payload that an SS may use to format and transmit a bandwidth request message or an initial ranging request message in a contention request zone. The value includes all PHY overheads as well as allowance for the MAC data the message may hold. It should be remembered that for OFDM and OFDMA PHYsical layers, a PS is defined as the duration of four OFDM symbols.

The BS always allocates bandwidth for contention IEs in integer multiples of these published individual transmission opportunity values. The number of transmission opportunities associated with a particular UL-MAP_IE corresponding to an initial ranging or bandwidth request interval is then dependent on the total size of this contention space allocation as well as the size of an individual transmission. See the numerical example in Section 10.3.7.

10.3.5.2 Contention Resolution Algorithm

Collisions may occur during initial ranging and bandwidth request intervals in the uplink (sub-) frame. The uplink transmission and contention resolution algorithm is the same for these two processes. Since an SS may have multiple active uplink service flows (and then, equivalently, multiple CIDs), it makes these ranging or request decisions on a per-CID or, equivalently, per-service QoS basis.

The method of contention resolution required by the 802.16 standard is based on a truncated binary exponential backoff, with the initial backoff window and the maximum backoff window values selected by the BS. These two parameters are specified in the UCD message. They are given as a power-of-two value −1 (minus 1). For example, a value of 4 indicates a backoff window between 0 and 15; a value of 10 indicates a backoff window between 0 and 1023. For these four windows, the range of values of n is 0–15; i.e. the possible sizes are between 0 and 65535.

The contention resolution algorithm works as follows. When an SS has information to send and wants to enter the contention resolution process, it sets its internal backoff window equal to the request (or ranging for initial ranging) initial backoff window defined in the UCD message. This UCD message is itself referenced by the UCD count in the UL-MAP message currently in effect.

The SS randomly selects a number within this backoff window. The obtained random value indicates the number of contention transmission opportunities that the SS will defer before transmitting. An SS considers only the contention transmission opportunities for which this transmission would have been eligible. The contention zones are defined in the standard as Request IEs (or Initial Ranging IEs for initial ranging) in the UL-MAP messages, identified by appropriate UL-MAP_IEs (see Section 10.3.7). Note that each IE may consist of more than one contention transmission opportunity. Using bandwidth requests as an example, consider an SS whose initial backoff window is 0–15 and assume it randomly selects the number 11. The SS must defer a total of 11 contention transmission opportunities. If the first available Request IE is for six requests, the SS will not use this Request IE and has five more opportunities to defer. If the next Request IE is for two requests, the SS has three more to defer. If the third Request IE is for eight requests, the SS transmits on the fourth opportunity, after deferring for three more opportunities (see Figure 10.6).

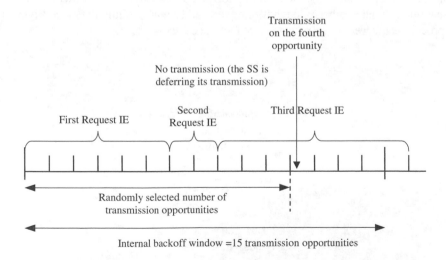

Figure 10.6 Example of a backoff mechanism. The SS has to wait 11 transmission opportunities (a randomly selected number between 0 and the internal backoff window). In this figure, only the Request IE (contention slot) is represented and not the rest of the uplink (sub-) frame

After a contention transmission, the SS waits for a Data Grant Burst Type IE in a subsequent map (or for a Ranging Response (RNG-RSP), message for initial ranging). Once received, the contention resolution is complete. For bandwidth requests, if the SS receives a unicast Request IE or Data Grant Burst Type IE at any time while deferring for this CID, it stops the contention resolution process and uses the explicit transmission opportunity. The SS considers the contention transmission lost if no data grant has been given within a given duration (or no ranging response within another given duration for initial ranging). In this case, the SS increases its backoff window by a factor of two, as long as it is less than the maximum backoff window. The SS then randomly selects a new number within its new backoff window and repeats the deferring process described above. This retry process continues until the maximum number of retries (i.e. Request Retries for bandwidth requests and Contention Ranging Retries for initial ranging) has been reached. At this time, for bandwidth requests, the SS discards the pending transmission. The minimum value for Request Retries is 16. Due to the possibility of collisions, bandwidth requests transmitted after a broadcast or multicast polling must be aggregate requests.

The choices of the Request (or Ranging) Backoff Start and the Request (or Ranging) Backoff End by the BS gives it much flexibility in controlling the contention resolution. These choices can be changed as frequently as the UCD message frequency if needed.

It is pointed out that this contention resolution algorithm is the same used for WiFi IEEE 802.11 WLAN for a contention-based distributed access function, which is the only mode effectively used, until now, for 802.11 WLANs.

10.3.6 Bandwidth Stealing

A bandwidth is always requested on a CID basis and allocated on an SS basis (to the SS basic CID). The process of bandwidth stealing is defined in the standard as the use, by a subscriber station (SS), of a portion of the bandwidth allocated in response to a Bandwidth Request for a connection to send another Bandwidth Request rather than sending data (see Figure 10.7). This process is allowed for some classes of QoS (see Chapter 11).

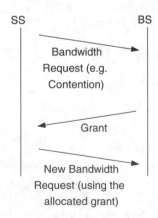

Figure 10.7   Illustration of the bandwidth stealing principle

10.3.7 Example of Uplink Access

The information sequence for unicast, multicast and broadcast polling is now illustrated in an example. The OFDM Layer is considered as well as the following numerical hypothesis:

Bandwidth = 3.5 MHz;
$n = 8/7$ (sampling factor);
G (guard time factor) = 1/8;
Frame duration = 5 ms;
Duplexing mode is FDD.

For these hypothesis, it can be verified that the number of OFDM symbols per frame is 69 (see Section 5.2.4). The standard states that an uplink subframe consists of a contention interval scheduled for initial ranging, contention interval(s) scheduled for Bandwidth Request purposes and one or multiple uplink PHY PDUs, each transmitted from a different SS.

Table 10.1 shows an example of UL-MAP MAC management message contents. The character of each IE, defined by an UL-MAP IE, changes depending on the type of CID used in this IE (see CID defined values in Table 7.1). When broadcast and multicast defined CID values are used, this is an invitation for all (or some of) the SSs to contend for requests. If a basic CID (then an SS's CID) is used, this is an invitation for a particular SS to transmit data and/or to request a bandwidth (see Section 10.3.2). In this table, two UL-MAP_IE fields, the subchannel index and the midamble repetition interval, are not shown in order to simplify the table. For OFDM (fixed WiMAX) PHYsical Layer parameters, i.e. UL-MAP_IE formats, UIUC values (see Table 9.9), etc., Start Time and Duration fields are in units of OFDM symbol duration (as for DL-MAP_IEs).

Initial ranging transmissions use a long preamble (two consecutive OFDM symbols) and the most robust mandatory burst profile. The most robust is BPSK with a Channel Coding rate of 1/2. It is estimated that the Ranging Request MAC management message is two OFDM symbols long. Then, the Initial Ranging Request PPDU is four OFDM symbols long. In the case of initial ranging, the maximum SS/BS round-trip propagation delay must also be taken into account. Four OFDM symbols are added (this is for a large cell).

The Request IE burst profile depends of the bandwidth request type as there are three possible uplink requests (see the UIUC table, Table 9.9). In this example, REQ Region Full is considered, which is the 'classical' uplink request. Thus, subchannelisation is not active. For these conditions, each transmit opportunity consists of a short preamble, i.e. one OFDM symbol and one OFDM symbol transmitting the bandwidth request, using the most robust mandatory burst profile. This symbol (96 uncoded bits) is enough to transmit the 48 bits of the MAC bandwidth request frame. The most robust burst profile is BPSK with a Reed–Solomon convolutional channel coding rate of 1/2. In fact, The Reed–Solomon convolutional coding rate of 1/2 is always used as the coding mode when requesting access to the network, except in subchannelisation modes, which use only convolutional coding of 1/2. Then, the Bandwidth Request PPDU is two OFDM symbols long.

The size of an individual transmission opportunity for each type of contention IE is indicated by the BS in the UCD MAC management message. This parameter, known as the bandwidth request opportunity size or the ranging request opportunity size, is the size in units of PS (Physical Slot) of the PHY payload that an SS may use to transmit a bandwidth request message or initial ranging request message in a contention request opportunity.

Table 10.1 Example of UL-MAP message contents. Two UL-MAP_IE fields, subchannel index and midamble repetition interval are not shown in this table

	UL-MAP message field(s)	Description
	Management message type of UL-MAP (=3)	
	Uplink channel ID (8 bits)	Identifier of the uplink channel to which this message refers (not to be confused with CID). Arbitrarily chosen by the BS, this ID acts as a local identifier for some transactions
	UCD count (8 bits)	Configuration change count of the UCD (the same as for the DL-MAP)
	Base Station ID (48 bits)	48-bit long field identifier of the BS (same as for the DL-MAP)
	Allocation start time (32 bits)	Effective start time of the uplink allocations defined by the UL-MAP, starting from the beginning of the downlink frame in which this UL-MAP message is placed
UL-MAP IE$_1$	CID = 0x0000 UIUC = 1 Start time = 0 Duration = 16	Defines the (Initial) Ranging IE
UL-MAP IE$_2$	CID = 0xFFFF UIUC = 2 Start time = 16 Duration = 12	Defines a (Bandwidth) Request IE associated with the broadcast CID. This is then a broadcast polling
UL-MAP IE$_3$	CID = 0xFF10 UIUC = 2 Start time = 28 Duration = 8	Defines a (Bandwidth) Request IE associated with CID = 0xFF10 (multicast CID, see the CID table, Table 7.1). This is then a multicast polling
UL-MAP IE$_4$	CID = 0xFF20 UIUC = 2 Start time = 36 Duration = 4	Defines a (Bandwidth) Request IE associated with CID = 0xFF20 (multicast CID, see the CID table, Table 7.1). This is then a multicast polling
UL-MAP IE$_5$	CID = 0x0023 UIUC = 5 Start time = 40 Duration = 10	Uplink grant (allocation) to CID = 0x0023 (the Basic CID of a specific SS). This corresponds to one uplink burst (or uplink PHY PDU), possibly containing more than one MAC message, transmitted in modulation/coding (in addition to other burst profile parameters) corresponding to UIUC = 5 (see the UIUC table, Table 9.9)
UL-MAP IE$_6$	CID = 0x0012 UIUC = 7 Start time = 50 Duration = 7	Uplink grant (allocation) to CID = 0x0012 (the Basic CID of a specific SS). UIUC = 7
UL-MAP IE$_7$	CID = 0x000A UIUC = 7 Start time = 57 Duration = 7	Uplink grant (allocation) to CID = 0x000A (the Basic CID of a specific SS). UIUC = 7
UL-MAP IE$_8$	CID = 0x0005 UIUC = 9 Start time = 64 Duration = 5	Uplink grant (allocation) to CID = 0x0005 (the Basic CID of a specific SS). UIUC = 9
UL-MAP IE$_9$	CID = 0 UIUC = 14 Start time = 69 Duration = 0	The end of the last allocated burst is indicated by allocating an End of Map burst (U IUC = 14, see the UIUC table, Table 9.9) with Duration field = 0 and CID = 0

Bandwidth Request IE = Bandwidth Request Contention Slots zone
(12 OFDM Symbols)

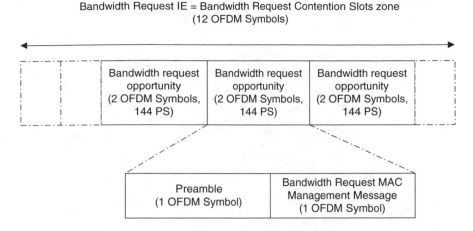

Figure 10.8 Example of Bandwidth Request IE (bandwidth request contention slots). The BS must allocate a bandwidth for Bandwidth Request IE in integer multiples of individual transmission opportunity values (indicated in the UCD message)

The PS is the basic unit of time. A PS corresponds to four (modulation) symbols used on the transmission channel (for OFDM and OFDMA PHY layers). The individual transmission opportunity (the bandwidth request opportunity size or the ranging request opportunity size) includes all PHY overheads as well as allowance for the MAC data the message may hold. It is assumed for this example that:

- The initial ranging request opportunity size, indicated by the UCD MAC management message, is equal to eight OFDM symbols. The initial ranging IE (contention slots), indicated by the UL-MAP MAC management message is 16 symbols long (see Table 10.1).
- The bandwidth request opportunity size, indicated by the UCD MAC management message, is equal to two OFDM symbols (see Figure 10.8). The Bandwidth Request IE (contention slots), indicated by the UL-MAP MAC management message, is 12 symbols long (see Table 10.1). These numerical values are used for the uplink frame figure.

It can be verified that the length of a bandwidth request opportunity size or the ranging request opportunity size (in PS, as indicated in the UCD message) is 144 PS (two OFDM symbols).

The duration field in UL-MAP_IE indicates the duration, in units of OFDM symbols, of the allocation. This value includes the preamble, the (possible) midambles and the postamble contained in the allocation. In the example given in this section, it is assumed that there is no midamble; i.e. the midamble repetition interval field (2 bits) in the beginning of the UL_MAP message (see Chapter 9 for the UL-MAP) is equal to 0b00. The standard indicates that all SSs must acquire and adjust their timing such that all uplink OFDM symbols arrive time coincident at the BS with an accuracy of ±50 % of the minimum guard interval or better. Figure 10.9 shows the uplink bursts in the uplink subframe described by the UL-MAP in Table 10.1.

Assuming that the Modulation and Coding Scheme (MCS) corresponding to UIUC = 7 is 16-QAM, 3/4, estimate the uplink data rate of the SS with Basic CID = 0x000A (on the duration of the considered frame)? Consider that this allocation is made once every four

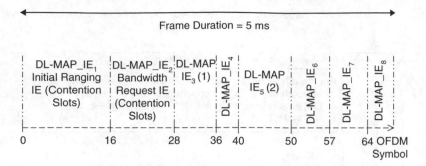

(1) Bandwidth Request IE associated with CID = 0xFF10 (multicast polling, contention slots
(2) Uplink grant (allocation) to CID = 0x0023 (the Basic CID of a specific SS)

Figure 10.9   Uplink bursts in the uplink subframe described in Table 10.1

frames (e.g. for an UGS class), what is the uplink data rate of the SS with Basic CID = 0 × 000A?

Data rate: UIUC = 7 → 16-QAM, 3/4;
7 symbols in 5 ms: $7 \times 192 \times 4 \times 3/4 = 4032$ bits in 5 ms → 806.5 kb/s;
One frame over four → 201.6 kb/s.

10.4 Contention-based Focused Bandwidth Request in OFDM PHY

The focused bandwidth or focused contention request in the OFDM PHYsical Layer is described in this section. This bandwidth request uses only part of all the OFDM subcarriers (instead of all of them as in the so-called full contention request) in association with contention codes. For the OFDM PHYsical Layer, two contention-based Bandwidth Request mechanisms, each referring to a REQ (Request) Region in the uplink frame, are defined in the standard:

• The full contention transmission corresponding to an uplink contention space called in the standard REQ Region Full, is indicated by a UIUC = 2 (see the UIUC table, Table 9.9), and is the contention mechanism used in Section 10.3.7 if subchannelisation is not activated.
• The focused contention transmission corresponding to an uplink contention space called in the standard REQ Region Focused, is indicated by a UIUC = 3 (see the UIUC table, Table 9.9). This transmission consists of a Contention Code modulated on a Contention Channel consisting of four OFDM subcarriers, both being randomly chosen by the candidate SS. The backoff procedure is always the one described in Section 10.3.5. The REQ Region Focused contention method has evidently a smaller collision probability than REQ Region Full.

The full contention transmission is mandatory. Capability of the focused contention transmission is optional. If the two types of request are possible, the SS may choose either of them.

10.4.1 Full Contention (REQ Region Full)

In a REQ Region Full:

- When subchannelisation is not active, the bandwidth request is the mechanism seen until this point in this book: each Transmission Opportunity (TO) consists of a short preamble (one OFDM symbol) and one OFDM symbol using the most robust mandatory burst profile (BPSK, coding rate of 1/2).
- When subchannelisation is active, the allocation is partitioned into TOs, both in frequency and in time. The width (in subchannels) and length (in OFDM symbols) of each TO are defined in the UCD message, along with the description of the burst profile corresponding to UIUC = 2. The number of subchannels used by each transmission opportunity may be 1, 2, 4, 8 or 16. The number of OFDM symbols that must be used by each transmission opportunity is also given in the UCD message. The transmission of an SS must contain a subchannelised preamble corresponding to the TO chosen, followed by data OFDM symbols using the most robust mandatory burst profile.

10.4.2 Focused Contention (REQ Region Focused)

In a REQ Region Focused, a transmission opportunity (sometimes called a transmit opportunity) consists of four subcarriers on two OFDM symbols (see Figure 10.10). Each transmission opportunity is indexed by consecutive transmission opportunity indices, the first occurring

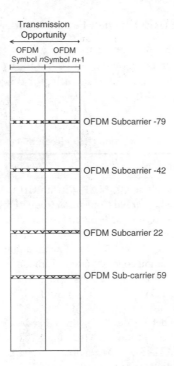

Figure 10.10 Example of the subcarriers of a focused contention transmission opportunity (contention channel index = 20). The SS transmits zero amplitude on all other subcarriers

transmission Opportunity being indexed 0. A candidate SS (requesting uplink bandwidth) sends a short code over a transmission opportunity as described below. This transmission is made in a REQ Region Focused defined by the BS using a UL-MAP_IE UIUC = 3 (see the UIUC table, Table 9.9).

The focused transmission consists of a contention code modulated on a contention channel consisting of four OFDM subcarriers. The SS transmits zero amplitude on all other subcarriers. The selection of the contention code is done with equal probability among eight possible codes of four bits each. The selection of the contention channel is done with equal probability among the time/frequency transmission opportunities that the concerned SS can use. There is no MAC message here; the BS only needs to detect a contention code on a contention channel. Upon detection, the BS provides an uplink allocation for the SS to transmit a Bandwidth Request MAC PDU and optionally additional data, but instead of indicating a Basic CID, a DL-MAP_IE is sent in combination with an OFDM Focused_Contention_IE (UIUC = 4), which specifies the contention channel, contention code and transmission opportunity that were used by the SS. This allows the SS to determine whether it has been given an allocation by matching these parameters with the parameters it used. The SS then can send a Bandwidth Request MAC PDU on the allocated uplink grant. This procedure is summarised in Table 10.2.

During a transmission opportunity, the amplitude of each of the four subcarriers must be boosted above its normal amplitude, i.e. the amplitude used during a noncontention OFDM symbol, including the current power-control correction. The boost in dB is equal to the value of the Focused Contention Power Boost parameter indicated by the BS in the UCD message.

10.4.2.1 Focused Contention (REQ Region Focused) with Subchannelisation

The number of contention codes that can be used by a Subchannelisation-enabled SS to request a subchannelised allocation is denoted C_{SE} in the standard. This value is given by the BS in the UCD message. The default value is 0 (typically, for BSs not supporting subchannelisation) and the allowed value range is 0–8.

Table 10.2 The steps of the focused contention (REQ Region Focused) procedure

Step		Action
0	The BS broadcasts	The BS sends an UL-MAP_IE with UIUC = 3, indicating a REQ Region Focused (Focused Contention Request region in the uplink frame). Other REQ-focused contention parameters are broadcasted on the UCD message
1	SS → BS	A candidate SS (an SS requesting an uplink bandwidth) sends a short code (a contention code, set of 4 bits, modulated on a contention channel, a set of four OFDM subcarriers) over a TO randomly chosen among the TOs (four OFDM subcarriers on two OFDM symbols) that this SS has the right to use.
2	BS → SS	If the BS detects the REQ-focused code, the BS provides an uplink allocation for the SS to transmit a Bandwidth Request in a (DL-MAP) OFDM Focused_Contention_IE (UIUC = 4), which specifies the contention channel, contention code and transmission opportunity that were used by the SS
3	SS → BS	The SS sends a Bandwidth Request MAC PDU on the allocated uplink grant

The contention code is selected randomly with equal probability from the appropriate subset of contention according to the value of C_{SE}. If the BS supports subchannelisation, only the last C_{SE} contention codes (among the eight available contention codes) may be used by subchannelisation-enabled SSs that wish to receive a subchannelised allocation. In response, the BS can:

- provide the requested allocation as a subchannelised allocation, where the UL-MAP IE for allocation of bandwidth in response to a subchannelised network entry signal, in the subchannelised section of the UL-MAP, is identified by UIUC = 13;
- provide the requested allocation as a full allocation (default);
- provide no allocation.

10.4.3 Summary of Contention-based Uplink Grant-request Methods

To end this section, Table 10.3 provides a summary of OFDM PHY contention-based uplink grant-request methods.

10.5 Contention-based CDMA Bandwidth Request in OFDMA PHY

The OFDMA PHY supports two mandatory contention-based Bandwidth Request mechanisms:

- The SS sends the bandwidth request header as specified for the OFDM Layer. The OFDM Layer focused contention and subchannelisation considerations no longer apply to OFDMA where subchannels (subcarriers) are distributed instead of OFDM symbols.
- Conversely, use the CDMA-based mechanism described below.

Table 10.3 Summary of OFDM PHY contention-based uplink grant-request methods

Full contention		Focused contention	
Transmitted in a REQ Region Full in the uplink frame		Transmitted in a REQ Region Focused in the uplink frame	
Without subchannelisation	With subchannelisation	Without subchannelisation	With subchannelisation
The basic method of contention-based uplink grant request (see the example in Section 10.3.7)	<ul><li>Allocation is partitioned into transmission opportunities (TOs) both in frequency and in time</li><li>The width (in subchannels) and length (in OFDM symbols) of each transmission opportunity (TO) are defined in the UCD message</li></ul>	See Table 10.2	<ul><li>The contention code sent by a candidate SS is selected randomly with equal probability from the appropriate subset of all contention codes according to the value of C_{SE}, indicated in UCD</li><li>In response, the BS may provide the requested allocation as a subchannelised allocation; provide a full allocation or provide no allocation</li></ul>

As specified in Chapter 9, the OFDMA PHY specifies a ranging subchannel. The BS provides in the UCD message a subset of ranging codes that are used for contention-based Bandwidth Requests and initial ranging. The BS can determine the purpose of the received code by the subset to which the code belongs. The requesting bandwidth SS selects, with equal probability, a ranging code from the code subset allocated to Bandwidth Requests. This ranging code modulates the ranging subchannel and is then transmitted during the appropriate uplink allocation. Upon detection, the BS provides an implementation-dependent uplink allocation for the SS, but instead of indicating a Basic CID, the broadcast CID is sent in combination with a CDMA_Allocation_IE, which specifies the transmit region and ranging code that were used by the SS. This allows an SS to determine whether it has been given an allocation by matching these parameters with the parameters it used. The SS then uses the allocation to transmit a Bandwidth Request MAC PDU and/or data.

If the Bandwidth Request does not result in a subsequent allocation of bandwidth, the SS assumes that the ranging code transmission resulted in a collision and then follows the contention resolution mechanism.

11

Network Entry and Quality of Service (QoS) Management

11.1 Ranging

The ranging process is defined in the 802.16 standard as the process that allows the SSs to:

- Acquire the correct timing offset of the network. The SS can then be aligned with the frame received from the BS (for OFDM and OFDMA PHYsical layers).
- Request power adjustments and/or downlink burst profile change. The SS can then be received within the appropriate reception thresholds.

The SS realises ranging by transmitting the RNG-REQ MAC management message. The RNG-REQ message may be sent in the initial ranging uplink contention slots (see Chapter 10) or in data grant intervals allocated by the BS to the SS. Ranging is made by the SS at initialisation and periodically. This gives two types of ranging processes:

- the initial ranging;
- the periodic ranging (sometimes simply known as ranging).

In the OFDM PHYsical Layer, initial ranging uses the initial ranging uplink contention slots while the periodic ranging uses the regular (allocated uplink grant) bursts. The initial ranging procedure is a part of the network entry process described in Section 11.6.

11.1.1 Ranging Messages

Two MAC management messages are used for initial ranging: the Ranging Request (RNG-REQ) message and the Ranging Response (RNG-RSP) message. These messages are now detailed.

The Ranging Request (RNG-REQ) message is transmitted by the SS at initialisation. It can also be used at other periods to determine network delay and to request power and/or downlink burst profile change. The CID field value in the MAC header of the RNG-REQ message contains the following:

- Initial ranging CID (0x0000) if the SS is attempting to join the network or if the SS has not yet registered and is changing downlink (or both downlink and uplink) channels.

Table 11.1   RNG-REQ message parameters. Some of these fields are TLV coded

RNG-REQ field	Description
Downlink channel ID	Identifier of the downlink channel on which the SS received the UCD describing the uplink channel on which this ranging request message is transmitted
Requested downlink burst profile	Indicated by the DIUC of the downlink burst profile requested by the the SS for downlink traffic
SS MAC address	Transmitted when the SS is attempting to join the network
MAC version	
Ranging anomalies	After the SS has received an RNG-RSP addressed to the SS, indicating that the SS is already at maximum power, minimum power, etc.
AAS broadcast capability	

- In all other cases, the Basic CID is used (other cases assume that a Basic CID was assigned to the SS).

The parameters that may be included in the RNG-REQ message are shown in Table 11.1.

The Ranging Response (RNG-RSP) message is transmitted by the BS in response to a received RNG-REQ. In addition, it may also be transmitted asynchronously to send corrections based on measurements that have been made on other received data or MAC messages. As a result, the SS is prepared to receive an RNG-RSP at any time, not just following an RNG-REQ transmission. Some of the parameters that may be included in the RNG-RSP message are shown in Table 11.2.

Two other MAC management messages are used for periodic ranging: the Downlink Burst Profile Change Request (DBPC-REQ) Message and the Downlink Burst Profile Change Response (DBPC-RSP) Message.

The Downlink Burst Profile Change Request (DBPC-REQ) message is sent by the SS to the BS on the SS Basic CID to request a change in the downlink burst profile used by the BS to transport data to the SS. The required burst profile for downlink traffic is indicated by its corresponding DIUC (as defined in the DCD message). The DBPC-REQ message is sent at the current (BS allocated) burst profile type. If the SS detects fading on the downlink, the SS uses this message to request transition to a more robust burst profile. The general format of the DBPC-REQ message is shown in Figure 11.1. A request for change of the downlink burst profile is made with the DBPC-REQ Message (instead of the RNG-REQ message as described above) if the SS has been granted an uplink bandwidth, i.e. a data grant allocation to the SS Basic CID. Otherwise, the SS uses the RNG-REQ message for a request for change of the downlink burst profile (see Section 11.1.3).

The Downlink Burst Profile Change Response (DBPC-RSP) message is transmitted by the BS on the SS Basic CID in response to a DBPC-REQ message from the SS. If the DIUC parameter is the same as requested in the DBPC-REQ message, the request was accepted. Otherwise, if the request is rejected, the DIUC parameter is the previous DIUC at which the SS was receiving downlink data. The general format is shown in Figure 11.2. The operation of DBPC Request and Response messages is illustrated in Figure 11.3.

Table 11.2 Some of the parameters of the RNG-RSP message. Some of these fields are TLV coded

RNG-RSP field	Description
Ranging status	Indicates whether uplink messages are received within acceptable limits by BS. This field is mandatory. Possible values are 1 = continue, 2 = abort, 3 = success, 4 = rerange
Timing adjust	The time required to advance the SS transmission to enable frames to arrive at the expected time instance at the BS. Units are PHYsical Layer specific. For the OFDM (WiMAX) PHYsical Layer, the unit is the symbol duration $1/f_s$). If this field is not included, no time adjustment is to be made
Power adjust	Transmitted power offset adjustment (signed 8-bit, in 0.25 dB units). This field specifies the relative change in the transmission power level that the SS must make in order for transmissions to arrive at the BS at the desired power. If this field is not included, no power adjustment is to be made
Offset frequency adjust	Transmitted frequency offset adjustment (signed 32-bit, Hz units). This field specifies the relative change in transmission frequency that the SS must make in order to better match the BS. This is a fine-frequency adjustment within a channel, not reassignment to a different channel
Basic CID	Assigned by the BS at initial access. In this case, the RNG-RSP message is being sent on the initial ranging CID in response to an RNG-REQ message that was itself sent on the initial ranging CID
Primary management CID	Assigned by the BS at initial access, in the same conditions as the Basic CID
MAC address	SS MAC address (48-bit)
Downlink operational burst profile	Specifies the least robust DIUC that may be used by the BS for transmissions to the SS
Frame number	In which the corresponding RNG-REQ message (for OFDM/ WiMAX) was received. When the frame number is included, the SS MAC address does not appear in the same message
Ranging subchannel	The OFDM ranging subchannel index that was used to transmit the initial ranging message
AAS broadcast permission	

Management message type (=23)	Reserved	DIUC	Configuration change count

Figure 11.1 DBPC-REQ MAC management message format. The required burst profile for downlink traffic is indicated by its corresponding DIUC (as defined in the DCD message)

Management message type (=24)	Reserved	DIUC	Configuration change count

Figure 11.2   DBPC-RSP MAC management message format. If the DIUC parameter is the same as requested in the DBPC-REQ message, the request was accepted

11.1.2 Initial Ranging

Initial ranging allows an SS joining the network to acquire the correct transmission parameters, such as time offset, frequency and transmitted power level, so that the SS can communicate with the BS.

In the OFDM PHYsical Layer, initial ranging uses the initial ranging uplink contention slots. First, an SS synchronises to the downlink using the preamble and then learns the uplink channel characteristics through the UCD MAC management message. Then, the SS scans the UL-MAP message to find an initial ranging (contention slots) interval. As described in Chapter 10, the BS allocates an initial ranging interval made of one or more (initial ranging) transmission opportunities. In this interval, the SS sends an RNG-REQ MAC management message, with a CID value = 0 (see Table 7.1). For the OFDMA PHY, the initial ranging process is different. It uses initial ranging CDMA codes (see Section 10.5). In OFDM PHY, initial ranging transmissions use a long preamble (two OFDM symbols) and the most robust mandatory burst profile.

When the initial ranging transmission opportunity occurs, the SS sends the RNG-REQ message (using a CDMA code in the case of the OFDMA PHY). The SS sends the message as if it were colocated with the BS, as the propagation delay is taken into account in the initial ranging transmission opportunity.

11.1.2.1 Initial Ranging Message Initial Transmitted Power Value

The SS calculates the maximum transmitted power for initial ranging, denoted $P_{TX\_IR\_MAX}$, as follows:

$$P_{TX\_IR\_MAX} = EIR x P_{IR,max} + BS\_EIRP - RSS$$

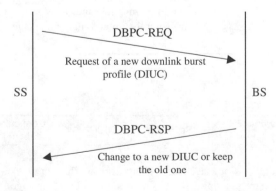

Figure 11.3   Illustration of DBPC Request and Response messages operation

where

- BS_EIRP (Equivalent Isotropic Radiated Power) is the BS transmitted power value transmitted in the DCD message.
- EIR$xP_{IR,max}$ is the maximum equivalent isotropic received power at the BS. This value is obtained from the DCD message.
- RSS is the measured Received Signal Strength Indicator (RSSI) at the SS.

It can be verified that the above equation is the realisation of:

Maximum transmitted power for initial ranging
= intended maximal received power at BS + estimated path loss between the SS and the BS.

The SS antenna gains may be included in the above formula. In the case that EIR$xP_{IR,max}$ and BS_EIRP are not known, the SS starts from the minimum transmit power level defined by the BS.

11.1.2.2 Successful Initial Ranging

The CIDs for the basic and primary management connections (see Section 8.4) are assigned in the RNG-RSP and REG-RSP messages. This ranging process is now described.

Once the BS has successfully received the RNG-REQ message, the BS returns a RNG-RSP message using the initial ranging CID. This RNG-RSP contains the MAC address of this new SS. Within the RNG-RSP message, the BS also puts the basic and primary management CIDs assigned to this SS. The same CID value is assigned to both members of each connection pair (uplink and downlink). The RNG-RSP message also contains information on the transmitted power level adjustment and offset frequency adjustment as well as any timing offset corrections. At this point the BS starts using individually allocated initial ranging intervals addressed to the SS Basic CID to complete the ranging process, unless the status of the RNG-RSP message is 'success', in which case the initial ranging procedure is finished. The RNG-REQ and RNG-RSP messages dialogues can also provide the CID value for the secondary management connection.

If the status of the RNG-RSP message is 'continue', the SS waits for an individual initial ranging interval assigned to its Basic CID. Using this interval, the SS transmits another RNG-REQ message using the Basic CID along with any power level and timing offset corrections. The BS sends another RNG-RSP message to the SS with any additional fine tuning required. The ranging request/response steps are repeated until the ranging response contains a ranging successful notification or the BS aborts ranging. Once successfully ranged (RNG-REQ is within tolerance of the BS), the SS joins normal data traffic in the uplink. This process is illustrated in Figure 11.4.

11.1.2.3 Unsuccessful Initial Ranging

If, after having sent the RNG-REQ message, the SS does not receive a response, it sends again the RNG-REQ message at another initial ranging transmission opportunity at one power level step higher. This step value is not fixed in the standard, although it indicates it cannot be greater than 1 dB. If the SS receives an RNG-RSP message containing the frame number in which its RNG-REQ message was transmitted, the SS considers that the transmission attempt was unsuccessful. This RNG-RSP message indicates that the BS has detected a transmission in the ranging slot that it is unable to decode. However, the SS implements the

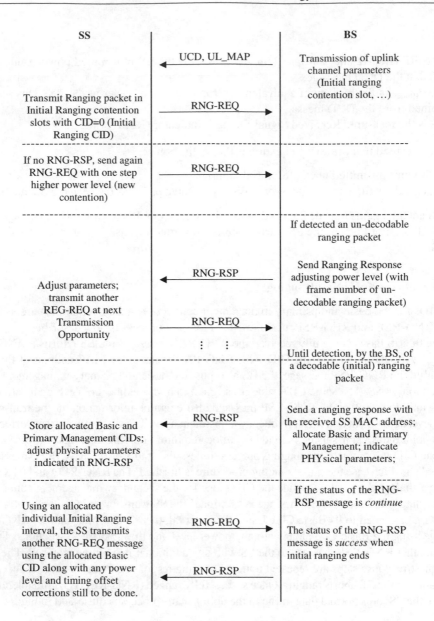

Figure 11.4   Illustration of the initial ranging process

corrections specified in the RNG-RSP message and issues another RNG-REQ message after the appropriate backoff delay.

11.1.3 Ranging (or Periodic Ranging)

After the initial ranging where physical parameters are adjusted, the periodic ranging allows the SSs to adjust transmission parameters so that the SSs can maintain communication quality with the BS. Distinct processes are used for managing the uplink and downlink. Some PHY modes support ranging mechanisms unique to their capabilities.

11.1.3.1 Downlink Ranging

In the downlink, if the received CINR goes outside an allowed operating region, according to the link adaptation mechanism, the SS requests a change to a new burst profile using one of two methods: the RNG-REQ message or the DBPC-REQ message. With both methods, the message is sent using the Basic CID of the SS:

- DBPC-REQ message. If the SS has been granted an uplink bandwidth, i.e. a data grant allocation to the SS Basic CID, the SS sends a DBPC-REQ message using that allocation. The BS responds with a DBPC-RSP message.
- RNG-REQ message. If a grant is not available and the SS requires a new burst profile on the downlink, the SS sends an RNG-REQ message in an initial ranging (contention slot) interval, using the same procedure as for initial ranging.

Link adaptation of the downlink is described in Section 11.2.

11.1.3.2 Uplink Ranging

In the uplink, periodic ranging is realised as follows. For each (unicast) uplink burst grant in which a signal is detected, the BS determines the quality of the uplink signal. If the signal is not within acceptable limits, the BS issues the RNG-RSP message including the appropriate correction data (see the RNG-RSP format above) and a status of 'continue'. If a sufficient number of correction messages are issued without the SS signal quality becoming acceptable, the BS sends the RNG-RSP message with a status of 'abort' and then terminates the link management of the SS. Accordingly, the SS processes the RNG-RSP messages it receives, implementing any PHYsical layer corrections that are specified (when the status is 'continue') or initiating a restart of MAC activities (when the status is 'abort').

The SS responds to each uplink bandwidth grant the BS addresses to it. When the status of the last RNG-RSP message received by the SS is 'continue', the SS includes the RNG-REQ message in the allocated transmitted burst. When the status of the last RNG-RSP message received is 'success' (due to the fact that the BS considers that the signal is now within acceptable limits), the SS uses the grant to service its pending uplink data queues. If no data is pending, the SS responds to the grant by transmitting a block of padded data.

For each (unicast) uplink burst grant, the BS determines whether or not a transmitted signal is present. If no signal is detected in a specified number of successive grants, the BS terminates the link management for the associated SS.

The possibility to change the burst profiles is the basis of the link adaptation mechanism, allowing a very efficient use of the radio resource. Link adaptation in 802.16/WiMAX is described in the following section.

11.2 Link Adaptation

Link adaptation has different applications in the downlink and in the uplink. The principle is always the same: choosing the most suitable Modulation and Coding Scheme (MCS) in order to have the highest possible data rate while fulfilling the CINR (quality) requirements.

11.2.1 Downlink Channel Link Adaptation

The downlink burst profile is determined by the BS according to the quality of the signal that is received by each SS. To reduce the volume of uplink traffic, the SS monitors the CINR and compares the average value against the allowed range of operation. This region is bounded by threshold levels indicated in the DCD message for each defined burst profile. If the received CINR goes outside the allowed operating region (see Chapter 9), the SS requests a change to a new burst profile using one of two methods: the RNG-REQ message or the DBPC-REQ message. The SS applies an algorithm to determine its optimal burst profile in accordance with the threshold parameters established in the DCD message. This algorithm is not specified in the standard and can be proposed by the vendor or the operator.

The messages exchanged between the SS and the BS for a burst profile change are not exactly the same whether an SS is moving to a more or less robust burst profile. Figure 11.5 shows the case where an SS is moving to a more robust type. Figure 11.6 shows a transition to a less robust burst profile.

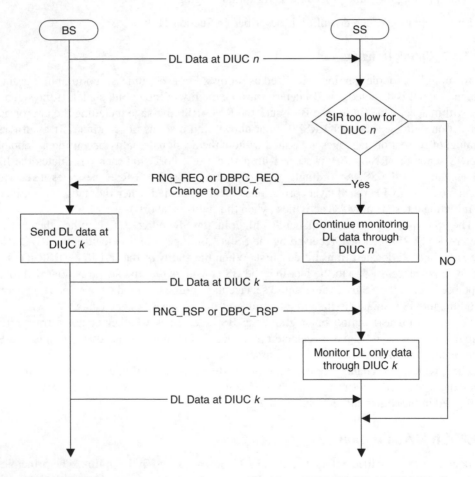

Figure 11.5   Transition to a more robust burst profile. (From IEEE Std 802.16-2004 [1]. Copyright IEEE 2004, IEEE. All rights reserved.)

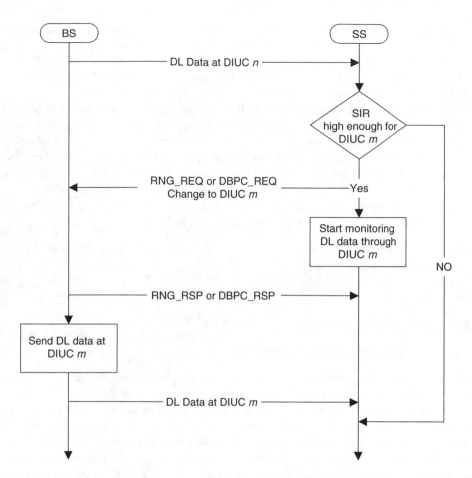

Figure 11.6 Transition to a less robust burst profile. (From IEEE Std 802.16-2004 [1]. Copyright IEEE 2004, IEEE. All rights reserved.)

11.2.2 Uplink Channel Link Adaptation

In the uplink, the burst profile is also (as for the downlink) decided by the BS. The RNG-RSP Message is used for that purpose as described in Section 11.1.3.

11.3 The Five Scheduling Services or QoS Classes

The IEEE 802.16 standard provides powerful tools in order to achieve different QoS constraints. The 802.16 standard MAC Layer provides QoS differentiation for the different types of applications that might operate over 802.16 networks, through five defined scheduling service types, also called QoS classes.

This classification into these scheduling service classes facilitates bandwidth sharing between different users. Every user has a quality of scheduling service class, also known as QoS class. According to this parameter, the BS scheduler allocates the necessary amount of bandwidth required for each application. This mechanism allows an efficient and adapted distribution of the existing resources. Therefore, a real-time application, such as a video application, will have the priority in bandwidth allocation in comparison with FTP (File Transfer Protocol) or email applications. This is not the case, for example, with the presently used WiFi (WLAN) system where all services have exactly the same level of QoS.

Scheduling services represent the data handling mechanisms supported by the MAC scheduler for data transport on a given connection. Uplink request (grant) scheduling is performed by the BS based on the scheduling service type, with the intent of providing each subordinate SS with a bandwidth for uplink transmissions and opportunities to request this bandwidth, when needed. As already mentioned in this book, each connection is associated with a single data service flow and each service flow is associated with a set of QoS parameters. These parameters are managed using the DSA and DSC MAC management messages dialogues (see Section 11.4). Four scheduling services were defined in 802.16e:

- Unsolicited Grant Service (UGS);
- real-time Polling Service (rtPS);
- non-real-time Polling Service (nrtPS);
- Best Effort (BE).

A fifth scheduling service type was added in 802.16e:

- Extended Real-time Polling Service (ertPS);

Each of these scheduling services has a mandatory set of QoS parameters that must be included in the service flow definition when the scheduling service is enabled for a service flow. The QoS parameters defined in the 802.16 standard are described in Section 7.4. Table 11.3 gives the mandatory service flow QoS parameters for each of the four scheduling services defined in 802.16-2004. If present, the minimum reserved traffic rate parameter of UGS must have the same value as the maximum sustained traffic rate parameter. Concerning ertPS, 802.16e indicates that the key service IEs are the maximum sustained traffic rate, the minimum reserved traffic rate, the maximum latency and the request/transmission policy.

Uplink request/grant scheduling is performed by the BS in order to provide each SS with a bandwidth for uplink transmissions and opportunities to request a bandwidth, when needed. By specifying a scheduling service and its associated QoS parameters, the BS scheduler can anticipate the throughput and latency needs of the uplink traffic and provide polls and/or

Table 11.3 Mandatory QoS parameters of the scheduling services defined in 802.16-2004. If present, the minimum reserved traffic rate parameter of the UGS must have the same value as the maximum sustained traffic rate parameter

Scheduling service	Maximum sustained traffic rate	Minimum reserved traffic rate	Request/ transmission policy	Tolerated jitter	Maximum latency	Traffic priority
UGS	•	(Can be present)	•	•	•	
rtPS	•	•	•		•	
nrtPS	•	•	•			•
BE	•		•			•

Table 11.4 Poll/grant options for each scheduling service

Scheduling service	Piggyback grant request	Bandwidth stealing	Unicast polling	Contention-based polling
UGS	Allowed	Allowed	PM (Poll-Me) bit can be used	Not allowed
ertPS	Extended piggyback	Allowed	Allowed	Allowed
rtPS	Not allowed	Not allowed	Allowed	Not allowed
nrtPS	Not allowed	Not allowed	Allowed	Allowed
BE	Not allowed	Not allowed	Allowed	Allowed

grants at the appropriate times. Table 11.4 summarises the poll/grant options available for each of the scheduling services.

More details for each scheduling service are provided in the following subsections.

11.3.1 Unsolicited Grant Service (UGS)

The UGS scheduling service type is designed to support real-time data streams consisting of fixed-size data packets issued at periodic intervals. This would be the case, for example, for T1/E1 classical PCM (Pulse Coded Modulation) phone signal transmission and Voice over IP without silence suppression.

In a UGS service, the BS provides fixed-size data grants at periodic intervals. This eliminates the overhead and latency of SS requests. Figure 11.7 illustrates the UGS mechanism. The BS provides Data Grant Burst IEs (UL-MAP_IEs, see Chapter 10) to the SS at periodic intervals based upon the maximum sustained traffic rate of the service flow. The size of these grants is sufficient to hold the fixed-length data associated with the service flow, taking into account the associated generic MAC header and grant management subheader.

The grant management subheader (see Chapter 10) is used to pass status information from the SS to the BS regarding the state of the UGS service flow. If the SI (Slip Indicator) bit of

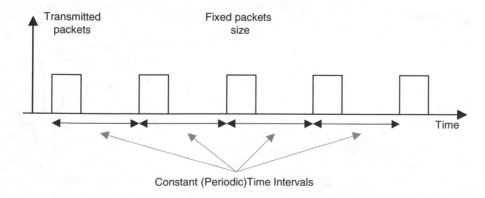

Figure 11.7 UGS scheduling service uplink grants allocation mechanism

the grant management field is set, the BS may grant up to 1 % additional bandwidth for clock rate mismatch compensation. The SSs that have an active UGS connection are not polled individually (by the BS) unless they set the PM bit in the header (precisely, in the grant management subheader) of a packet on the UGS connection.

11.3.2 Extended Real-Time Polling Service (ertPS)

The ertPS (extended real-time Polling Service) class was added by the 802.16e amendment. The standard [2] indicates that ertPS is a scheduling mechanism that builds on the efficiency of both UGS and rtPS. The BS provides unicast grants in an unsolicited manner like in UGS, thus saving the latency of a bandwidth request. However, whereas UGS allocations are fixed in size, ertPS allocations are dynamic. The ertPS is suitable for variable rate real-time applications that have data rate and delay requirements. An example is Voice over IP without silence suppression.

11.3.3 Real-Time Polling Service (rtPS)

The rtPS scheduling service type is designed to support real-time data streams consisting of variable-sized data packets that are issued at periodic intervals. This would be the case, for example, for MPEG (Moving Pictures Experts Group) video transmission.

In this service, the BS provides periodic unicast (uplink) request opportunities, which meet the flow's real-time needs and allow the SS to specify the size of the desired grant. This service requires more request overheads than UGS, but supports variable grant sizes for optimum real-time data transport efficiency. Figure 11.8 shows the rtPS mechanism.

11.3.4 Non-Real-Time Polling Service (nrtPS)

The nrtPS is designed to support delay-tolerant data streams consisting of variable-size data packets for which a minimum data rate is required. The standard considers that this would be the case, for example, for an FTP transmission. In the nrtPS scheduling service, the BS provides unicast uplink request polls on a 'regular' basis, which guarantees that the service flow receives request opportunities even during network congestion. The standard states that the BS

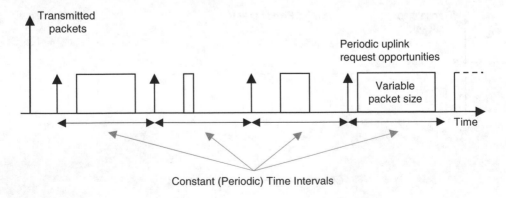

Figure 11.8   rtPS scheduling service uplink grants allocation and request mechanism

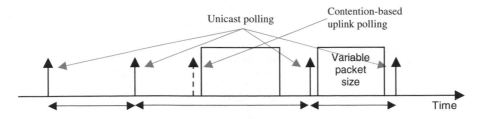

Regular (not necessarily periodic) time intervals

Figure 11.9   Illustration of the nrtPS scheduling service uplink grants allocation and request mechanism. The SS may use contention request opportunities as well as unicast request opportunities

typically polls nrtPS CIDs on an interval on the order of one second or less. In addition, the SS is allowed to use contention request opportunities, i.e. the SS may use contention request opportunities as well as unicast request opportunities. Figure 11.9 shows the nrtPS mechanism.

11.3.5 Best Effort (BE)

The BE service is designed to support data streams for which no minimum service guarantees are required and therefore may be handled on a best available basis. The SS may use contention request opportunities as well as unicast request opportunities when the BS sends any. The BS do not have any unicast uplink request polling obligation for BE SSs. Therefore, a long period can run without transmitting any BE packets, typically when the network is in the congestion state. Figure 11.10 shows the BE mechanism.

11.4 Scheduling and Deployment of Services Over WiMAX

11.4.1 The Scheduler is in the BS!

As already mentioned in this book, two topologies are defined: Point to MultiPoint (PMP) and Mesh. In the PMP mode, the network operates with a central BS and probably with a

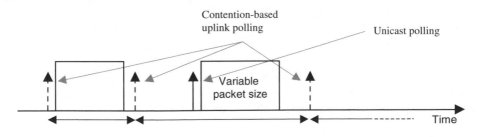

Completely nondeterministic time intervals

Figure 11.10   Illustration of the BE scheduling service uplink grants allocation and request mechanism. The BS does not have any unicast uplink request polling obligation for a BE SS

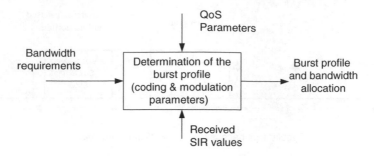

Figure 11.11   The BS decides for bandwidth and burst profile allocations according to many entry parameters

sectorised antenna that is capable of handling multiple independent sectors simultaneously. WiMAX/802.16 uses the PMP centralised MAC architecture where the BS scheduler controls all the system parameters (radio interface). It is the role of the BS scheduler to determine the burst profile and the transmission periods for each connection; the choice of the coding and modulation parameters are decisions that are taken by the BS scheduler according to the quality of the link and the network load and demand. Therefore, the BS scheduler must permanently monitor the received CINR values (of the different links) and then determine the bandwidth requirements of each station taking into consideration the service class for this connection and the quantity of traffic required. Figure 11.11 shows the BS scheduler operation.

By specifying a scheduling service and its associated QoS parameters, the BS scheduler can anticipate the throughput and latency needs of the uplink traffic. This is a mandatory operation in determining the appropriate burst profile for each connection. The BS may transmit without having to coordinate with other BSs, except possibly for the Time Division Duplexing (TDD) mode, which may divide time into uplink and downlink transmission periods common for different BSs.

Based on the uplink requests and taking into account QoS parameters and scheduling services priorities, the BS scheduler decides for uplink allocations. These decisions are transmitted to the SSs through the UL-MAP MAC management message. Figure 11.12 shows the BS scheduler operation for the uplink. The BS scheduler also decides for the downlink and transmits the decision using the DL-MAP MAC management message. Figure 11.13 shows the BS scheduler operation for the downlink.

There is also a scheduler present in the subscriber station (SS). The role of this scheduler is to classify all the incoming packets into the SS different connections.

The standard does not define a scheduling algorithm that must be used. Any of the known scheduling algorithms can be used: Round Robin, Weighted Round Robin, Weighted Fair Queuing and probably other known or to be defined scheduling algorithms. New scheduling algorithms are already being proposed specifically for WIMAX/802.16 scheduling in the literature.

11.4.2 Scheduling of the Different Transmission Services

Each SS to BS (uplink) connection is assigned a scheduling service type as part of its creation. When packets are classified in the Convergence Sublayer (CS), the connection into

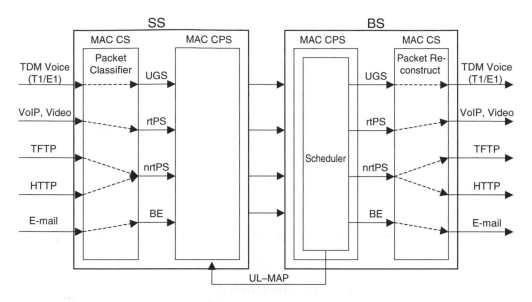

Figure 11.12   BS scheduler operation for the uplink [5]

which they are placed is chosen based on the type of QoS guarantees that are required by the application (see Figures 11.12 and 11.13).

Although the standard gives all the details about the different classes of QoS and the methods of bandwidth allocation, the details of scheduling and reservation management are left unstandardised and are then left for vendors and operators. In Table 11.5 the scheduling

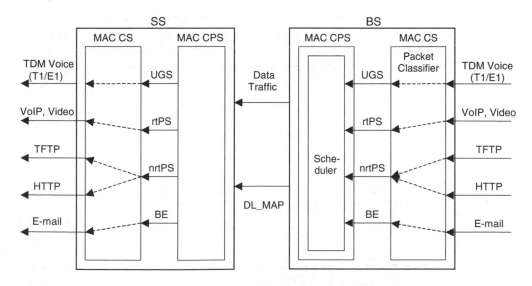

Figure 11.13   BS scheduler operation for the downlink [5]

Table 11.5   Scheduling service type (or QoS class) for some services.

Application	Expected class of QoS	Explicitly indicated by the standard
T1/E1	UGS	Yes
VoIP without silence suppression	UGS	Yes
VoIP with silence suppression	ertPS	Yes
MPEG	rtPS	Yes
FTP	nrtPS	Yes
TFTP	nrtPS	No
HTTP	nrtPS	No
Email	BE	No

service type is given that can be used for some classical services. Some of these services are mentioned in the standard.

11.5 Dynamic Service Addition and Change

11.5.1 Service Flow Provisioning and Activation

A service flow that has a non-Null ActiveQoSParamSet is said to be an active service flow (see Section 7.2.2). This service flow may request and be granted a bandwidth for the transport of data packets. An admitted service flow may be activated by providing an ActiveQoS-ParamSet, signalling the resources desired at the current time.

A service flow may be provisioned and then activated. Alternatively, a service flow may be created dynamically and immediately activated (see Figure 11.14). In this latter case, the two-phase activation is skipped and the service flow is available for immediate use upon authorisation.

The provisioning of service flows is outside the scope of the 802.16 standard. This should be part of the network management system. During provisioning, a service flow is classified, given a 'provisioned' flow type and a service flow ID. Enabling service flows follow the transfer of the operational parameters. In this case, the service flow type may change

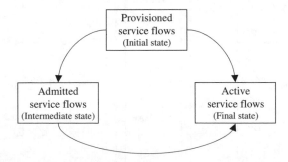

Figure 11.14   Possible transitions between service flows. A BS may choose to activate a provisioned service flow directly or may choose to take the path to active service flows passing by the admitted service flows

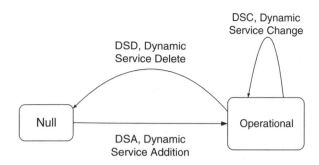

Figure 11.15   Dynamic service flow operations. (Based on Reference [1].)

to 'admitted' or to 'active'; in the latter case, the service flow is mapped on to a certain connection.

Service flows may be created, changed or deleted. This is accomplished through a series of MAC management messages:

- DSA (Dynamic Service Addition) messages create a new service flow;
- DSC (Dynamic Service Change) messages change an existing service flow;
- DSD (Dynamic Service Delete) messages delete an existing service flow. This is illustrated in Figure 11.15.

For some service flows, it may be specified that the DSA (Dynamic Service Addition) procedure used for service flow creation must be activated by the network entry procedure. Triggers other than network entry may also cause creation, admission or activation of service flows. These triggers are said to be outside the scope of the standard. Service flow encodings contain either a full definition of service attributes or a service class name. A service class name must be an ASCII string, which is known at the BS and which indirectly specifies a set of QoS parameters.

11.5.2 Service Flow Creation

Creation of a service flow may be initiated by either the BS (mandatory capability) or the SS (optional capability). The DSA messages are used to create a new service flow. Since it is a new service flow, the primary management CID is used to establish it. This CID value is used in the generic MAC header of DSA messages.

A DSA-REQ, DSA REQuest MAC management message from an SS, wishing to create either an uplink or downlink service flow, contains a service flow reference and a QoS parameter set, marked either for admission-only or for admission and activation. A DSA-REQ from a BS contains an SFID for either one uplink or one downlink service flow, possibly its associated CID, and a set of active or admitted QoS parameters. In both cases, the BS checks successively the following points:

- whether the SS is authorised for service;
- whether the service flow(s) QoS can be supported;

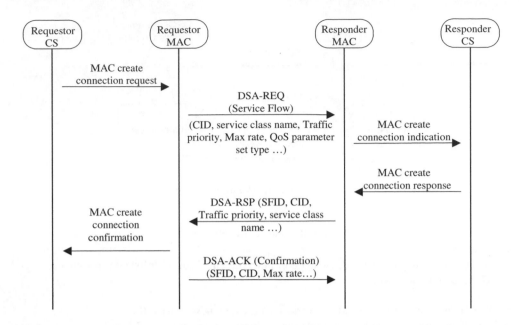

Figure 11.16 Successful service flow creation procedure messages and attributes. Some of the parameters in this figure are not included in some DSx messages depending on whether the service creation is BS or SS initiated. (Figure from Reference [1] modified by G. Assaf.)

and then possibly creates SFID and, if AdmittedQoSParamSet is non-null, maps the service flow to a CID. The BS or the SS responds with the DSA-RSP, DSA Response, MAC management message, indicating acceptance or rejection. Figure 11.16 shows the service flow creation procedure messages.

For DSA-REQ and DSC-REQ messages sent by an SS, the DSX-RVD (DSx Received) message is generated by the BS to inform the SS that the BS has correctly received the DSx-REQ message. This can be done quickly before sending the DSx-RSP message, which is transmitted only after the DSx-REQ is authenticated.

The general format of DSA-REQ, DSA-RSP and DSA-ACK messages is shown in Figure 11.17. The DSA-REQ message cannot contain parameters for more than one service flow. The transaction ID is a unique identifier for this transaction assigned by the sender. The confirmation code indicates the status for the dynamic service (DSx-xxx) messages. Possible values are: OK-success, reject, reject-service-flow-exists, reject-header-suppression, reject-authentication-failure, etc.

Management message type (=11, 12 or 13)	Transaction ID	Confirmation code (only DSA RSP and ACK)	TLV encoded information

Figure 11.17 The general format of DSA-REQ, DSA-RSP and DSA-ACK MAC management messages

The TLV encoded information field contains the specification of the service flow's traffic characteristics, CS specific parameters and scheduling requirements. These parameters are:

- CID: specifies the CID assigned by the BS to a service flow. The CID is used in bandwidth requests and in MAC PDU headers.
- SFID: the primary reference of a service flow. Only the BS may issue an SFID in BS-initiated DSA-REQ/DSC-REQ messages and in its DSA-RSP/DSC-RSP to SS-initiated DSA-REQ/DSC-REQ messages. The SS specifies the SFID of a service flow using this parameter in a DSC-REQ message.
- Service class name: see Section 11.5.1 above.
- QoS parameter set type: provisioned, admitted or active.
- Traffic priority. The value of this parameter specifies the priority assigned to a service flow. Given two service flows identical in all QoS parameters besides priority, the higher priority service flow should be given lower delay and higher buffering preference. For nonidentical service flows, the priority parameter should not take precedence over any conflicting service flow QoS parameter. No specific algorithm for using this parameter is given in the standard. For uplink service flows, the BS uses this parameter when determining precedence in request service and grant generation.
- Scheduling service type. This is the one that should be enabled for the associated service flow, between the five defined scheduling service types: BE (default), nrtPS, rtPS, ertPS or UGS. If this parameter is omitted, BE service is assumed. An undefined scheduling service type (implementation-dependent) can also be set.
- Other QoS parameters: maximum sustained traffic rate, maximum traffic burst, minimum reserved traffic rate, minimum tolerable traffic rate, ARQ parameters for ARQ-enabled connections, etc. (see QoS parameters in Section 7.4).
- Target SAID (Security Association IDentifier). This indicates the SAID on to which the service flow that is being set up will be mapped. The SAID is a security association identifier shared between the BS and the SS (see Chapter 15).
- CS specification. This specifies the CS that the connection being set up will use. Possible choices are No CS, Packet IPv4, Packet IPv6, Packet 802.3/Ethernet, Packet 802.1Q VLAN, Packet IPv4 over 802.3/Ethernet, Packet IPv6 over 802.3/Ethernet, Packet IPv4 over 802.1Q VLAN, Packet IPv6 over 802.1Q VLAN and ATM.

11.5.3 Service Flow Modification and Deletion

Both provisioned and dynamically created service flows are modified with the DSC message, which can change the admitted and active QoS parameter sets of the flow. A single DSC message exchange can modify the parameters of either one downlink service flow or one uplink service flow.

A successful DSC transaction changes the service flow QoS parameters by replacing both the admitted and active QoS parameter sets. If the message contains only the admitted set, the active set is set to null and the flow is deactivated. If the message contains neither set, then both sets are set to null and the flow is de-admitted. When the message contains both QoS parameter sets, the admitted set is checked first and, if the admission control succeeds, the active set in the message is checked against the admitted set in the message to ensure that it is a subset. If all checks are successful, the QoS parameter sets in the message become the new admitted and active QoS parameter sets for the service flow. If either of the checks fails, the DSC transaction fails and the service flow QoS parameter sets are unchanged. Some service flow parameters, including the service flow scheduling type, may not be changed with the DSC messages.

An SS wishing to delete a service flow generates the DSD-REQuest message to the BS. The BS verifies that the SS is really the service flow 'owner' and then removes the service flow. The BS responds using the DSD-RSP message. On the other hand, a BS wishing to delete a dynamic service flow, no longer needed, generates a delete request to the associated SS using a DSD-REQuest. The SS removes the service flow and generates a response using a DSD-RSP.

11.5.4 Authorisation Module

The authorisation module is a logical function within the BS that approves or denies every change to QoS parameters and classifiers associated with a service flow. This includes every DSA-REQ message aiming to create a new service flow and every DSC-REQ message aiming to change a QoS parameter set of an existing service flow. Such changes include requesting an admission control decision (e.g. setting the AdmittedQoSParamSet) and requesting activation of a service flow (e.g. setting the ActiveQoSParamSet).

In the static authorisation model, the authorisation module stores the provisioned status of all deferred service flows. Admission and activation requests for these provisioned service flows are permitted as long as the admitted QoS parameter set is a subset of the provisioned QoS parameter set and the active QoS parameter set is a subset of the admitted QoS parameter set. Requests to change the provisioned QoS parameter set are refused. Requests to create new dynamic service flows are refused. This defines a static system where all possible services are defined in the initial configuration of each SS.

In the dynamic authorisation model, the authorisation module communicates through a separate interface to an independent policy server. This policy server may provide the authorisation module with advance notice of upcoming admission and activation requests, and it specifies the proper authorisation action to be taken on those requests. Admission and activation requests from an SS are then checked by the authorisation module to ensure that the ActiveQoSParamSet being requested is a subset of the set provided by the policy server. Admission and activation requests from an SS that are signalled in advance by the external policy server are permitted. Admission and activation requests from an SS that are not presignalled by the external policy server may result in a real-time query to the policy server or may be refused.

Prior to the initial connection setup, the BS retrieves the provisioned QoS set for an SS. This is handed to the authorisation module within the BS. The BS caches the provisioned QoS parameter set and uses this information to authorise dynamic flows that are a subset of the provisioned QoS parameter set. The standard states that the BS should implement mechanisms for overriding this automated approval process (such as described in the dynamic authorisation model). For example it could:

(a) deny all requests whether or not they have been pre-provisioned;
(b) define an internal table with a richer policy mechanism but seeded by the provisioned QoS set;
(c) refer all requests to an external policy server.

11.6 Network Entry

Systems must follow a list of procedures for entering and registering a new SS or, more generally, a new node to the network. The network entry procedures described in this section apply only to the PMP topology. The network entry procedure for Mesh (not included in WiMAX profiles for the moment) is described in the standard.

Figure 11.18 shows an illustration of the network entry sequence of procedures. The sequence of procedures for network entry is described in the following:

1. Scan for a downlink channel and establish synchronisation with the BS. On initialisation or after signal loss, the SS must acquire a downlink channel. The SS has nonvolatile storage

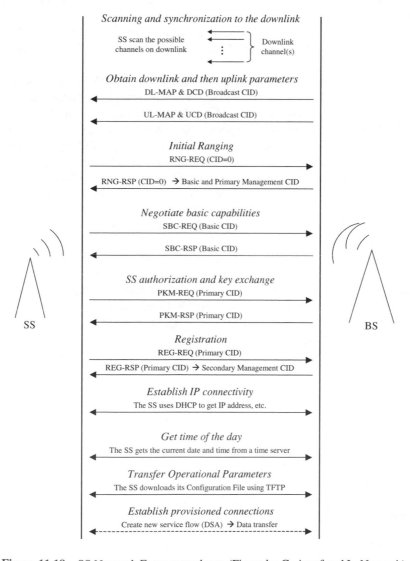

Figure 11.18 SS Network Entry procedures. (Figure by G. Assaf and L. Nuaymi.)

in which the last operational parameters are stored. It first tries to reacquire this downlink channel. If this fails, the SS begins to scan the possible channels of the downlink frequency band of operation continuously until it finds a valid downlink signal.

The frame start preamble is a repetitive well-known pattern and the SS may use it in order to synchronise timing and frequency parameters with the BS. The BS may further employ active scanning or diversity methods to speed up and enhance the process of downlink synchronisation. At this stage, the SS needs the DL-MAP, DCD, UCD and UL-MAP messages.

2. Synchronisation. The SS MAC searches for the DL-MAP message. The SS achieves MAC synchronisation once it has received at least one DL-MAP message. An SS MAC remains in synchronisation as long as it continues to successfully receive the DL-MAP and DCD messages of the downlink channel. If the standard-defined Lost DL-MAP interval is elapsed without a valid DL-MAP message, an SS must try to reestablish synchronisation. The Lost DL-MAP interval (denoted T21 in the standard) is equal to 11 s, as updated in 802.16e.

Obtain transmit (uplink) parameters from the UCD message. After synchronisation, the SS waits for a UCD message from the BS in order to obtain the set of transmission parameters needed for a possible uplink channel. UCD and uplink transmission parameters (UL-MAP) messages are transmitted periodically by the BS (see Chapters 9 and 10).

3. Perform initial ranging. The main objective of initial ranging is the adjustment of each SS timing offset and power parameters in the initialisation phase. Initial ranging (see Section 11.1 at the beginning of this chapter for details of this procedure) takes place when an SS wishes to enter a network. The SS sends an RNG-REQ message in a contention-based initial ranging interval. The CID field is set to the noninitialized SS value (initial ranging CID=0).

The CIDs for the basic and primary management connections are assigned during this initial ranging procedure (see Section 11.1.2).

4. Negotiate basic capabilities. This is the phase where SS and BS exchange their supported parameters. After completion of ranging, the SS informs the BS of its basic capabilities by transmitting an SBC-REQ (SS Basic Capability Request) message with its capabilities. The SBC-REQ message is transmitted by the SS during the initialisation (Network Entry) phase to inform the BS of its basic capabilities. The following parameters are included in the basic capabilities request:

- Bandwidth allocation: support for full frequency duplexing if each of the H-FDD and FDD is supported (for a TDD profile there is no alternative).
- Physical parameters: maximum transmit power (for each of the four possible modulations: BPSK, QPSK, 16-QAM and 64-QAM), current transmit power (used for the burst that carries the SBC-REQ Message), focused contention support (OFDM PHY specific), SS demodulator support (64-QAM, CTC, BTC, STC and AAS), SS modulator support (64-QAM, BTC, CTC, subchannelisation and focused contention BW request), SS focused contention support (see Chapter 10) and SS TC sublayer support, the FFT size (OFDMA PHY specific), permutations support, MIMO parameters support, AAS parameters support, security parameters support, power control parameters support, power save parameters support, handover parameters support, etc.
- Size of FSN (Fragment Sequence Number) values used when forming MAC PDUs on non-ARQ connections, such as the ability to receive requests piggybacked with data, etc.

The BS responds with an SBC-RSP (SS Basic Capability Response) message with the intersection of the SS and BS capabilities. The SBC-RSP message is transmitted by the BS in response to a received SBC-REQ. Its role is to confirm or not the proposed parameters in the SBC-REQ. If the BS does not recognise an SS capability, it may return this as 'off' in the SBC-RSP.

5. Authorise the SS and perform keys exchange. Next, the SS has to exchange secure keys which is part of the authentication mechanism. This is realised through the PKM protocol. The SS sends a PKM-REQ (Privacy Key Management Request) message to the BS. The BS responds with a PKM-RSP (Privacy Key Management Response) message. This exchange is detailed in Chapter 15.

6. Perform registration. Registration is the process by which the SS is allowed entry into the network and, specifically, a managed SS receives its secondary management CID and thus becomes a manageable SS. This part of the Network Entry process is detailed in Section 11.6.1 below.

Basic MAC and primary MAC management connections are established during the SS initial ranging (see above). These connections are not secure. A secondary management connection is established when the authorisation procedure is finished during SS registration. This connection is used for the IP connectivity establishment and for the Trivial File Transfer Protocol (TFTP) file configuration loading (see the following steps). The secondary management connection is secure. Figure 11.19 illustrates the sequence between initial ranging and registration of the SS Network Entry process.

Implementation of phases 7, 8 and 9 at the SS is optional. These phases have to be performed only if the SS has indicated in the REG-REQ message that it is a managed SS.

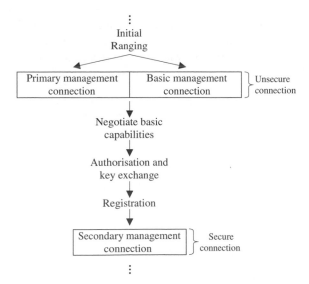

Figure 11.19 Initial ranging until registration part of the SS Network Entry process. This figure shows the case of a managed SS, i.e. having a secondary management connection

7. Establish IP connectivity. At this point, the SS uses the Dynamic Host Configuration Protocol (DHCP) [17,18] mechanisms in order to obtain an IP address, from the DHCP server
   and any other parameters needed to establish IP connectivity. If the SS has a configuration
   file (containing, for example, tables of QoS filters), the DHCP response contains the name
   of a file that contains further configuration parameters. The IP parameters of the SS are set
   up based on the DHCP server response. Establishment of IP connectivity is performed on
   the SS secondary management connection. The secondary management messages are carried in IP datagrams (see Section 5.2.6 of the standard for IP CS PDU formats).
8. Establish the time of day. The SS needs to have the current date and time from the BS.
   This is required for time-stamping logged events. It can be needed, for example, for some
   encryption algorithms. Accuracy is to the nearest second. The protocol by which the SS
   retrieves the time of day from a time server through the BS (no authentication is needed)
   is defined in IETF RFC 868 [19], which gives the number of seconds starting from year
   1900, on 4 bytes. The request and response are transferred using the User Datagram
   Protocol (UDP).
       The time retrieved from the server, the Universal Coordinated Time (UTC), must be
   combined with the time offset received from the DHCP response to create the SS current
   local time. Establishment of the time of day is performed on the SS secondary management connection.
9. Transfer operational parameters. After the DHCP procedure is successful, the SS down-
   loads its configuration file using the Trivial File Transfer Protocol (TFTP) [20] on the SS
   secondary management connection, as shown in Figure 11.20, if specified in the DHCP
   response (the TFTP configuration file server is specified in the DHCP response). The TFTP
   is a rather simple protocol used to transfer files, working over the UDP.
       When the configuration file download has been completed successfully, the SS notifies
   the BS by transmitting a TFTP-CPLT (Config File TFTP Complete Message) MAC management message on the SS primary management connection. Transmissions of TFTP-
   CPLT messages by the SS continue periodically until a TFTP-RSP (Config File TFTP
   Complete Response) MAC management message is received with an 'OK' response from
   the BS or the SS terminates retransmission due to retry exhaustion. The SS configuration
   file includes many system informations such as boot information.

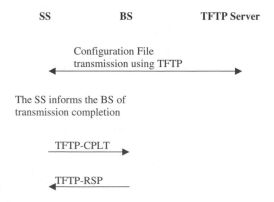

Figure 11.20   Transfer of the SS configuration file (operational procedure)

10. Set up connections. After the transfer of operational parameters (for a managed SS) or after registration (for an unmanaged SS), the BS sends DSA-REQ messages to the SS to set up connections for preprovisioned service flows belonging to the SS. The SS responds with DSA-RSP messages as detailed in Section 11.3 above.

11.6.1 Registration

Registration is then the process by which the SS is allowed entry into the network and, specifically, a managed SS receives its secondary management CID and thus becomes a manageable SS. To register with a BS, the SS sends the REG-REQ MAC management message to the BS. The general format of the REG-REQ message is shown in Figure 11.21.

In the REG-REQ message, the SS indicates the following:

- ARQ Parameters: fragmentation and ARQ parameters applied during the establishment of the secondary management connection.
- SS management support: whether or not the SS is managed by standard-based IP messages over the secondary management connection; if 'yes', this a so-called managed SS.
- IP management mode. This dictates whether the provider intends to manage the SS on an ongoing basis via IP-based mechanisms.
- IP version. This indicates the version of IP used on the secondary management connection: 4(default) or 6. When the SS includes the IP version in the REG-REQ, the BS includes the IP version parameter in its REG-RSP. The BS decides the use of one of the IP versions supported by the SS.
- SS capabilities encodings: ARQ support (indicates the availability of SS support for ARQ), MAC CRC support (indicates whether or not the SS supports the MAC level CRC), multicast polling group CID support (indicates the maximum number of simultaneous multicast polling groups the SS is capable of belonging to), authorisation policy support (indicates whether the SS can apply the IEEE 802.16 security, constituting X.509 digital certificates and the RSA public key encryption algorithm, as authorisation policy), etc.
- Vendor ID encoding: vendor identification.

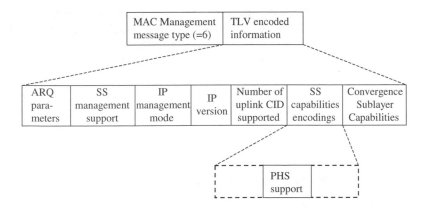

Figure 11.21   General format of the REG-REQ message. Not all possible TLV encodings are represented in this figure

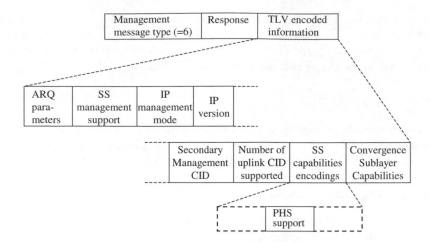

Figure 11.22 General format of the REG-RSP message. Not all possible TLV encodings are represented in this figure

- MAC version encoding.
- maximum number of supported security associations: specifies the maximum number of supported security associations of the SS.
- Convergence Sublayer (CS) capabilities (Classification/PHS options: ATM, Packet IPv4 or v6, etc.; by default, Packet, IPv4 and 802.3/Ethernet should be supported, level of PHS support).

The BS responds with a REG-RSP (REGistration RESponse) message. For an SS that has indicated being a managed SS in its REG-REQ message, the REG-RSP message includes the BS-allocated secondary management CID. The general format of the REG-RSP message is shown in Figure 11.22. The response field indicates message authentication success or failure. Message authentication verification is based on HMAC or CMAC (see Chapter 15 for message authentication).

In this message, the BS indicates the mode of SS management operation, the MAC version used in the network, Vendor ID Encoding of the BS, response to the REG-REQ indication of whether or not the requester wishes to accept IP-based traffic on the secondary management connection, once the initialisation process has been completed. Also included in the REG-RSP is a response to the capabilities of the requesting SS provided in the REG-REQ (if the request included capabilities information) if they can be used. If a capability is not recognised, the response indicates that this capability will not be used by the requesting SS. Capabilities returned in the REG-RSP cannot be set to require greater capability of the requesting SS than indicated in the REG-REQ.

11.6.2 De-registration and Re-registration

The DREG-CMD (De/Re-register Command) MAC management message is transmitted by the BS to force the SS to change its access state: stop using the current channel, use it with restrictions, return to normal mode, etc. The DREG-CMD can be unsolicited or in response to an SS DREG-REQ message. The DREG-REQ SS de-registration MAC management is sent by the SS to the BS in order to notify the BS of the SS de-registration request from the BS and the network.

11.6.3 SS Reset

The BS may transmit the RES-CMD (Reset Command) MAC management message on an SS Basic CID to force this SS to reset itself, reinitialise its MAC and then repeat initial system access. This message may be used if an SS does not respond to the BS or if the BS detects continued abnormalities in the transmissions of an SS.

The main parameter of RES-CMD is the message authentication. This is done with the HMAC key sequence number concatenated with an HMAC-Digest (see Chapter 15).

Part Four

Diverse Topics

12

Efficient Use of Radio Resources

With the contribution of Jérôme Brouet, Alcatel, France

12.1 Introduction

WiMAX systems are designed to provide high spectral efficiency in order to offer very high data rates to users. The robust physical layer based on the OFDM/OFDMA technique and the use of high-performance coding are partly responsible for the resulting high spectral efficiency (see Chapters 5 and 6).

On the one hand, spectral efficiency is enhanced by WiMAX Radio Resource Management (RRM) mechanisms (see Section 12.3). Those mechanisms, specified in the IEEE 802.16 standards include admission control, power control, link adaptation and dynamic frequency selection [1,2]. In addition to these radio mechanisms, the WiMAX Forum defines a framework and a functional split of the radio resource management procedures in the WiMAX radio access network (WiMAX architecture is described in Chapter 13). This framework actually further optimise RRM procedures (mainly for admission control and handover) and is especially needed for consistent RRM operations in a multivendor WiMAX radio network.

On the other hand, high spectral efficiency and performance superiority over the other radio systems is also achieved by the use of advanced antenna technology systems. The IEEE 802.16 standards and associated WiMAX profiles are presently defining and taking provisions for several alternatives for implementing advanced antenna technologies (see Section 12.4). In the present version of a mobile WiMAX, both adaptive antenna technology with beamforming and MIMO technology are possible. While multiple antenna technologies had only been adopted in enhanced releases for 3GPP and 3GPP2 standards, which makes their massive deployment challenging in an existing mature network, these technologies are actually supported from the very first release of mobile WiMAX systems, making them a real feature for first WiMAX radio network deployments and consequently the key cornerstone for excellent radio performance.

Whereas previous techniques are clearly improving the performance of the systems for unicast connections between a BS and an MS (Mobile Station), the WiMAX Forum has also taken provision for an efficient support of Multicast/Broadcast Services (MBS) over the WiMAX air interface. In order to achieve high performance, WiMAX system deployment should follow some guidelines exposed hereafter.

WiMAX: Technology for Broadband Wireless Access   Loutfi Nuaymi
© 2007 John Wiley & Sons, Ltd

12.2 Radio Engineering Consideration for WiMAX Systems

WiMAX solutions include fixed WiMAX and mobile WiMAX. According to the type of terminal offered (outdoor CPE (Consumer Premise Equipment), self-install indoor CPE, mobile terminal–PCMCIA, laptop, smart phones), each solution has a different deployment method.

12.2.1 LOS/NLOS Propagation

Coverage from a BS is linked to several radio parameters. The first one is the propagation environment. Depending on the relative locations of the BS and the terminal, several models can be used to evaluate the losses due to propagation.

12.2.1.1 LOS and Near LOS Propagation

Line-of-Sight (LoS) and Near Line-of-Sight (NLoS) propagations may happen when the BS and the MS are deployed outdoor, above the average height of the environment. This may be the case of the deployment of a fixed WiMAX solution in a rural environment with the BS located on a high altitude point (or at the top of a mast) and the SS (then a CPE, for a fixed WiMAX) deployed on the rooftop of the customer's house.

LOS propagation requires that the first Fresnel zone is free from any obstacle. In that case, the propagation losses are proportional to the square of the distance between the BS and the SS [22]. However, in a practical deployment, this happens very seldom. Usually, the Fresnel zone is obstructed and/or there are few obstacles on the BS to the CPE transmission path. In that case, near LOS models are used. However, in the case where there are few obstacles (a building, a water tower, a hill), the obstacles are modelled by knife-edge. An example of such a model can be found in Reference [23].

12.2.1.2 NLOS Propagation

NLOS (Non-Line-of-Sight) propagation occurs when the terminal is located indoor and/or at ground level. In this situation, there is in most cases no direct path between the BS and the terminal, there is a high number of obstacles on the BS to MS path (buildings, trees, cars, etc.) and the receiver may receive several copies of signal that experienced several reflections/diffractions on different obstacles. This type of propagation is typical of cellular deployments.

In the case of WiMAX, NLOS propagation corresponds to a deployment of a fixed or mobile WiMAX using self-install indoor terminals (wireless DSL deployment), or a deployment of a mobile WiMAX with mobile terminals (PCMCIA, laptop with integrated chipset, multimode mobile phones, etc.).

In NLOS, the losses versus the distance are much higher that in LOS. Usually, the distance decay exponent is between 3 and 4. An example of a propagation model that can be used to evaluate propagation losses in NLoS is the Erceg model [24].

12.2.2 Radio Parameters and System Gains

In order to evaluate the range, additional parameters are required. First, the value of some propagation parameters depends on additional radio parameters such as:

- Frequency band. WiMAX could be deployed at different frequencies (2.3 GHz, 2.5 GHz and 3.5 GHz); the higher the frequency, the higher the propagation losses.
- BS antenna height. Propagation losses decrease if the antenna height is increased.
- Terminal antenna height. The lower the terminal, the higher the losses (for mobile deployment, the height of the terminal is usually taken to be between 1.5 and 2 m).

Then, depending on the deployment scenario, a margin modelling the fluctuations of the radio environment needs to be considered.

In the case of the LOS/near LOS scenario, the BS and the MS are fixed. However, the propagation loss fluctuates (according, for example, to geoclimatic parameters). Hence, to evaluate the range, a margin needs to be evaluated according to the signal availability required (e.g. 99 %) and the geoclimatic parameters. A method to evaluate the margin is available in Reference [25].

In the case of the NLOS scenario, in order to reflect the distribution of the obstacles in the coverage area, a margin, called the shadowing margin, needs to be taken into account for the evaluation of the range. This shadowing effect is modelled by a lognormal distribution with a standard deviation, which depends on the environment (typically from 10 dB in an urban environment to 5 dB in rural environments). The resulting margin also depends on the signal probability availability on the cell area (usually between 90 and 95 %). A method to evaluate this margin can be found in Reference [22]. In addition, other margins may be included in the propagation losses: indoor margin (in the case of indoor coverage requirements), interference margin (to reflect the interference level generated by other cells transmitting at the same frequency), body losses, etc.

Finally, to evaluate the range, the system gain may be evaluated. The system gain is the maximum signal level difference that can be accepted on the BS-to-terminal path so that the receiver may decode a signal with sufficient quality. The system gains depend on many specific parameters:

- Transmission power of the BS/terminal.
- Antenna gain of the BS/terminal. The gain at the terminal side depends on the type of terminal. Outdoor CPEs may have gains in excess of 14 dBi while mobile terminals may have only a 0 dBi antenna gain.
- Receiver sensitivity. The receiver sensitivity depends on the signal bandwidth, the modulation and the coding scheme. The IEEE 802.16 standards are providing minimum reference values but vendors may come with better performance solutions.

12.2.3 WiMAX Radio Features that Enhance the Range

For fixed WiMAX solutions deploying outdoor CPEs, the coverage may be of several km (see Reference [26]). However, to get range figures in line with an exiting cellular system for WiMAX cellular-like deployments, additional features are needed to compensate for the extra losses due to lower terminal antenna gains, indoor penetration and NLOS behaviour.

The radio enhancement feature applicable to fixed and mobile WiMAXs is subchannelisation. This is the possibility to concentrate the transmit power on a few subchannels and thus to play with trade-off between the coverage and the maximum data rate a terminal can get at the cell edge. Other enhancement features that are only applicable to the mobile WiMAX are:

- Convolutional Turbo Coding (CTC).
- Repetition Coding. For each retransmission, up to 3 dB gain can be obtained. This again allows a trade-off between the maximum range and data rate at the cell edge.
- Hybrid ARQ (HARQ). This is the capability of the receiver to combine several transmissions of the same MAC PDUs.
- Advanced antenna technologies: beamforming and MIMO.

12.2.4 Frequency Planning Guidelines

Again, according to the deployment scenario, the frequency planning guidelines are different.

12.2.4.1 Fixed WiMAX

In the case of a fixed WiMAX with an outdoor CPE deployed in LOS/near LOS, similar frequency planning to that of Fixed Wireless Access (FWA) systems (e.g. PMP microwave transmission) can be used. At the radio site, several alternatives for sector configurations may be employed (1, 3, 4 and 6 sectors per site). In FWA-like deployment, the CPE has an external antenna with high gain and hence a reduced antenna beam width. Consequently, the antenna of the CPE provides significant interference protection; hence in order to achieve the CINR target on the service area, frequency planning with 3 and 4 frequencies may be used. For the fixed WiMAX with an indoor self-install CPE, which also have directive antennas, a similar frequency plan may be used.

For densification, more sectors (up to six) or more frequencies per sector can be deployed on the existing sides.

12.2.4.2 Mobile WiMAX

In the case of a mobile WiMAX deployment with mobile terminals, there is no interference reduction due to the MS antenna (omnidirectional antennas). However, because of the use of OFDMA and permutations (which gives some randomness in the use of subcarriers), and because of the better radio performance (more coding capabilities, beamforming and/or MIMO), it is possible to deploy an efficient system with a frequency planning reuse scheme of 3 and even 1 [10].

In the case of MBS (see below), the frequency channel assigned for transmitting multicast/broadcast connections must be deployed in a Single Frequency Network (SFN) manner.

For densification, the same techniques used for existing cellular systems are employed: more frequencies per site, cell splitting, microcellular layer, indoor coverage from the indoor, etc.

12.2.5 Base Station Synchronisation

IEEE 802.16-2004 indicates that there are three options of BS synchronisation [1]:

- Asynchronous configuration. Every BS uses its own permutation. The frame lengths and starting times are not synchronised among the base stations. This configuration can be used as an independent low-cost hot-spot deployment.
- Synchronous configuration. All the BSs use the same reference clock (e.g. by using GPS, or Global Positioning System). The frame durations and starting times are also synchronised

among the BSs but each BS may use different permutations. Due to the time synchronisation in this scenario and the long symbol duration of the OFDMA symbol, fast handovers as well as soft handovers are possible. This configuration can be used as an independent BS deployment with a controlled interference level.

- Coordinated synchronous configuration. All the BSs work in the synchronous mode and use the same permutations. An upper layer is responsible for the handling of subchannel allocations within the sectors of the base station, making sure that better handling of the bandwidth is achieved and enabling the system to handle and balance loads between the sectors and within the system.

The standard indicates that, for TDD and FDD realisations, it is recommended (but not required) that all BSs should be time-synchronised to a common timing signal.

12.3 Radio Resource Management Procedures

12.3.1 Power Control

The support of power control procedure is mandatory in the uplink. The procedure includes both an initial calibration process and a periodic update process.

By measuring the received power at the BS side, the BS can send power offset indications to the MS, which adjusts its transmit power level accordingly. It has to be noted that in the case where the MS uses only a portion of the subchannels, the power density remains unchanged regardless of the number of subchannels actually used (unless the maximum power level is reached).

The power offset indication sent by the BS is consistent with the actual MCS that is to be used for the uplink transmission. Power adjustments related to the use of a different modulation scheme is then taken into account. Finally, the power offset is also consistent with the maximum power that the MS can transmit (indicated in the SBC-REQ MAC message).

After reception of the power offset information, the MS adjusts its transmit power according to Equation (12.1) by simply adding the offset to the last value used for transmission:

$$P_{New} = P_{Last} + \text{Offset} \tag{12.1}$$

where P_{last} is the last used transmit power, P_{new} the new transmit power and 'Offset' the accumulation of offset values sent by the BS since the last transmission. Transmission power offset information is sent in RNG-RSP messages in units of 0.25 dB. This process allows for power fluctuations at a rate of at most 30 dB/s with a depth of at least 10 dB.

In the case of OFDMA-based WiMAX systems, the power control differs for some uplink burst types. For uplink burst regions used for the fast feedback channel (UIUC = 0), CDMA ranging (UIUC = 12) and CDMA allocation IE (UIUC = 14), the transmit power update formula of the MS is the following:

$$P_{New} = P_{Last} + (C/N_{new} - C/N_{last}) - 10[\log(R_{new}) - \log(R_{last})] + \text{Offset} \tag{12.2}$$

where
P_{new}, P_{last} and 'Offset' are as defined in Equation (12.1),
C/N_{new} is the normalised C/N (Carrier over Noise) of the new MCS used in the region,
C/N_{last} is the normalised C/N of the last used MCS,
R_{new} is the number of repetitions of the new MCS used in the region,
R_{last} is the number of repetitions of the last MCS.

Table 12.1 Normalised C/N values for power control procedures for OFDMA-based WiMAX terminals. (From IEEE Std 802.16e-2005 [2]. Copyright IEEE 2006, IEEE. All rights reserved.)

MCS	Normalised C/N (dB)
ACK region	-3
Fast feedback region	0
CDMA code	3
QPSK 1/3	0.5
QPSK 1/2	6
QPSK 2/3	7.5
QPSK 3/4	9
16-QAM 1/2	12
16-QAM 2/3	14.5
16-QAM 3/4	15
16-QAM 5/6	17.5
64-QAM 1/2	18
64-QAM 2/3	20
64-QAM 3/4	21
64-QAM 5/6	23

The normalised C/N values are default values given by the standard (see Table 12.1). However, the BS may override these values using a dedicated UCD message TLV.

The power control mechanism may also be implemented in the downlink (by, for example, limiting the interference created to the other cells). However, its implementation, if any, is vendor-specific.

12.3.2 Dynamic Frequency Selection (DFS)

The DFS mechanisms may be required in the case of deployment of a WiMAX system in a license-exempt band (e.g. the 5.8 GHz band). In that case, the BS and the MS implement a set of mechanisms that permits:

- sounding the radio environment prior to the use of a channel;
- periodically detecting 'specific spectrum users' (a specific spectrum user is a user that has been identified by the regulator as requiring strict protection from harmful interference);
- discontinuing operation on a channel after detection of a specific spectrum user;
- scheduling of periodic sounding testing periods (from the BS and the MS),
- selecting/changing to a new channel.

In any case, the BS cannot use a channel without testing the channel for other users, including specific spectrum users.

An example of a simplified process flow for DFS operation is depicted in Figure 12.1. At initialisation, the BS sounds the channels based on predefined timing parameters. Before potentially using a new channel, the BS must sound this channel for at least a 'startup test period' against other users. After completing the scanning of the channels, the BS can choose to operate on a channel that is not used by specific spectrum users and can then establish a connection with the

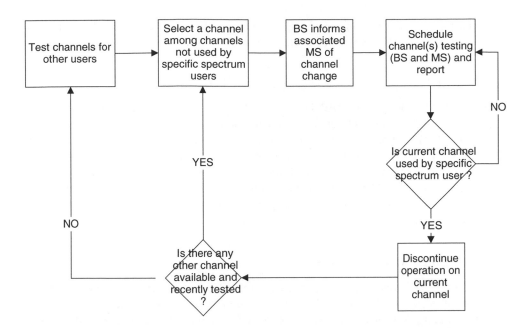

Figure 12.1   Example of a simplified process flowchart for a WiMAX system implementing DFS

MS in its coverage area. The mechanisms to sound the channels, to measure the interference, to detect specific spectrum users and to select a channel for operation are left to the manufacturer.

During the operation on a channel, the BS may be assisted by the MS to measure the use of one or more channels by other users. The scheduling of this process can be done either during a quiet period in the cell or during normal operation. When the BS requests MS support, the BS informs the MS by a channel measurement IE in the DL-MAP. Then, if the channel to be measured is the operating channel, the BS suspends all transmission/scheduling of the MAC PDU to any MS in the cell area during the measurement interval. During the testing process, the MS stores several parameters: the frame number corresponding to the first measurement, the accumulated time of measurement and the existence of a specific spectrum user in the channel. Those parameters are reported back to the BS during a measurement report response. However, if the MS detects a specific spectrum user, it will send an unsolicited REP-RSP message to the BS as soon as possible.

Upon detection of a specific spectrum user on the operating channel (either by the BS or the MS), the BS must discontinue the transmission of data MAC PDUs and MAC management message MAC PDUs within predefined periods ('Max Data Operation Period' and 'Management Operation Period' respectively).

When the operation needs to be discontinued, the BS starts to select a new channel for operation either from a recent and valid tested channel set or by resuming a similar process to that used at initialisation. After the selection of a new channel, the BS informs its associated MSs about channel change by including in the DCD message the new channel number and the frame number where the switch occurs.

In the case of DFS, the regulatory authority defines the timers involved in the sounding procedure channel (during initial sounding or periodic sounding) as well as the thresholds in order to prevent harmful interference to other users.

Finally, a similar mechanism can be implemented for other purposes than DFS and may be applied to any WiMAX deployments in shared channel environments.

12.3.3 Other Radio Resource Management Procedures

To optimise the performance of IEEE 802.16-based systems, other radio resource management procedures are implemented. The admission control of new connections is part of the RRM operation. The WiMAX Forum defines a framework to support and optimise the admission control (see Section 12.3.5). Admission control decision algorithms are left to the manufacturers.

Link adaptation mechanisms are also implemented. Again, the way the process operates and the selection of the MCS according to channel conditions and other local criteria are left to the vendor. More details on the supporting primitives for link adaptation are provided in Chapter 11.

12.3.4 Channel Measurements

In order for the BS to take appropriate decisions for radio resource management (power control, selection of the modulation and coding scheme and use of advanced antenna technology), the standard defines a set of channel quality indicators. Two families of indicators are available:

- RSSI (Received Signal Strength Indicator), which gives information on the received power level;
- CINR (Carrier-to-Interference-and-Noise Ratio), which gives information on received carrier-to-interference levels.

WiMAX radio equipment (MS and BS) can implement the means to measure, compute and report these indicators. In addition, to reflect the fluctuations of the radio channel in time, two statistics of the indicators are evaluated and reported: the mean and standard deviation.

12.3.4.1 Received Signal Strength Indicator (RSSI)

RSSIs are derived from measured received power level samples. The reported RSSIs are an average (in linear scale, e.g. in mW) of the measured power level samples (averaging is done by an exponential filter with a forgetting factor provided as a configuration parameter by the BS).

The mean RSSI is obtained from

$$\mu_{RSSI}[k] = \begin{cases} R[0] & \text{if } k = 0 \\ (1 - \alpha_{avg})\mu_{RSSI}[k-1] + \alpha_{avg}R[k] & \text{if } k > 0 \end{cases} \quad (mW) \qquad (12.3)$$

where $R[k]$ is the measured power sample during message k and α_{avg} is the averaging factor. The sample index k is incremented for every frame and the power is measured over the frame preamble. The averaging factor is transmitted by the BS either in a DCD message TLV or in a REP-REQ MAC message. Otherwise, the default value of 1/4 should be considered by the MS.

The standard deviation of RSSI is derived from the expectation-square statistics of the measured signal levels, x^2_{RSSI}, defined by

$$x^2_{RSSI}[k] = \begin{cases} |R[0]^2| & \text{if } k = 0 \\ (1-\alpha_{avg})x^2_{RSSI}[k-1]+\alpha_{avg}|R[k]^2| & \text{if } k > 0 \end{cases} \quad (\text{mW}^2) \qquad (12.4)$$

The method to measure the received signal strength is vendor-specific. However, the measurements should remain inside a ±4 dB absolute error.

The reported values are sent in MAC REP-REQ message using the dBm (dB) scale for the mean RSSI (respectively the standard deviation), as specified by

$$\mu_{RSSI\ dBm}[k] = 10\log(\mu_{RSSI}[k]) \qquad (\text{dBm})$$
$$\sigma_{RSSI\ dBm}[k] = 5\log(|x^2_{RSSI}[k]-(\mu_{RSSI}[k])^2|) \qquad (\text{dB}) \qquad (12.5)$$

The values are quantized with a 1 dB increment and each is sent in a 1-byte field in the MAC REP-REQ message. The range of RSSI spans from −123 dBm to −40 dBm.

12.3.4.2 Carrier-to-Interference and Noise Ratio (CINR)

For the IEEE 802.16-2004-based systems (using OFDM), the reported CINR indicators are physical CINR indicators. When the BS requests a CINR measurement report from the MS, the MS will answer back by including in the REP-RSP MAC message the estimates of the average and standard deviations of the CINR. CINR values are reported in a dB scale with 1 dB steps, ranging from −10 dB to 53 dB.

Reported CINR values are averaged using the same averaging method as that for the RSSI. The method to evaluate the CINR is vendor-specific. The measurement samples can be taken either from detected or pilot samples.

For the IEEE 802.16-2005-based systems (using OFDMA), additional options are specified. On the one hand, different CINR measurements are defined: physical and effective CINR measurements. On the other hand, the CINR reports may be sent either through the REP-REQ MAC message (mean and/or standard deviation of CINR) or through the fast-feedback channel (mean CINR only).

Several physical CINR reports can be requested from the MS:

- Physical CINR measurements on the preamble. In the case of frequency reuse 3 networks, the CINR is measured over modulated carriers of the preamble, while in the case of the frequency reuse 1 network, the CINR are measured over all the subcarriers (modulated or not, excluding the guard bands and DC channel).
- Physical CINR measurement from a permutation zone. In this case, CINR samples are measured from the pilots in the permutation zone.

In addition, the BS may also request 'effective CINR' measurements from a permutation zone on the pilot subcarriers. The effective CINR is a function of the physical CINR, taking into account channel conditions and implementation margins (implementation is manufacturer-dependent).

With this option, the MS has the additional possibility of indicating the MCS that best fits the specified target error rate to the BS.

12.3.5 Support of Radio Resource Management in the WiMAX RAN

12.3.5.1 RRM Functional Spit in the WiMAX RAN

The WiMAX Forum also defines a framework to optimise the operation of RRM in a WiMAX radio access network (also called an Access Service Network (ASN), see Chapter 13), and also supports multivendor interoperability of RRM procedures in the long term.

As specified by the WiMAX Forum, the RRM function in the network may provide reporting facilities and decision support to several network functions, such as:

- admission control, e.g. ensuring that enough radio resource is available at the BS side to serve appropriately a new MS or connection or service flow (either at service request or after a handover);
- handover preparation and control, e.g. optimising the choice of the target BS according to radio and BS load indicators.

Optionally, the RRM may also be involved in transport network resource management.

This framework is based on a functional split of the RRM functions into two parts (Radio Resource Agent (RRA) and Radio Resource Controller (RRC)) that communicate through standardised primitives [21].

The primitives exchanged are used either to report information (from RRA to RRC or between RRCs) or to communication decision support information (from RRC to RRA). This information includes measurement reports per MS and spare capacity per BS.

12.3.5.2 Future Enhancements of the RRM Network Function

As of today, the BS autonomously and independently performs the power control and interference management procedures. However, the RRM framework defined by the WiMAX Forum leaves the door open for further enhancement of the RRM procedures. Possible enhancement could be done by exchanging additional information (e.g. on channel configurations) between RRM entities in order to have a global optimisation of radio resource and not only a BS per BS optimisation.

12.4 Advanced Antenna Technologies in WiMAX

To improve radio performance, antenna technologies are very often used in cellular systems. A very popular solution in a wireless transmission system is to use receiver diversity at the BS side. The signals transmitted by the terminal are received at the BS by multiple antennas (usually two or four) and the signals from the different received paths are combined. A very popular technique for combining these different signals is the Maximum Ratio Combining (MRC) technique, which combines (or weights) the same symbol received from each branch according to their reception quality. The outcome is an increase in the receiver sensibility at the BS and consequently a range extension and/or the possibility to use a less robust radio transmission mode for a higher transmission rate. The order of magnitude of the improvement is of the order of the diversity order, e.g. for a system with two receiver antennas, the gain is typically about 3 dB.

The received diversity scheme is very efficient if the signals coming from the different antennas are uncorrelated. The correlation of the antenna mainly depends on the distance between the antennas (usually, a separation of 10 to 20 λ is required in a macrocellular environment): this is called spatial diversity. An operational alternative to such antenna configurations is to use cross-polarised antennas; this limits the visual impact on the environment.

The implementation and support of simple receive diversity in a BS is vendor-specific and does not require any standard mechanisms. Presently, more advanced antenna techniques also exist: smart antenna technology with beamforming and MIMO antennas. Each technology has its own advantages and system deployment constraints are developed further in the next subsections.

Also, both technologies require the support of the standard in order to get the full benefits of their operation. The IEEE 802.16 standards, especially the IEEE 802.16e amendment, provide all the hooks for the support of both antenna technologies.

12.4.1 Beamforming or AAS Technologies

Beamforming technologies may be encountered behind several wordings: smart antenna, beamforming and Adaptive Antenna System (AAS). In the following beamforming will be used.

12.4.1.1 Beamforming Basics

The main objective of beamforming technology is to take benefit from the space/time nature of the propagation channel. Indeed, due to multiple reflections, diffraction and scattering on the transmitter to receiver path in a cellular environment, the energy reaching the BS comes from multiple directions, each direction being affected by a different attenuation and phase.

In a macrocellular environment (i.e. the antenna of the BS is above the rooftop) the signals reaching the BS are inside a cone. The angular spread of the signal depends on the environment. In a urban environment, the angular spread is of the order of 20 degrees. In a more open environment, like in a rural environment, the angular spread is a few degrees.

In the uplink, the beamforming technology principle is to coherently combine the signals received for N antenna elements of an antenna array. A generic beamforming diagram is shown in Figure 12.2. A block diagram of a beamforming receiver (respectively transmitter)

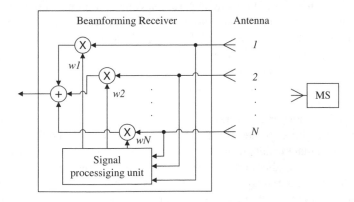

Figure 12.2   Example of a block diagram of a beamforming receiver with an N-element antenna array

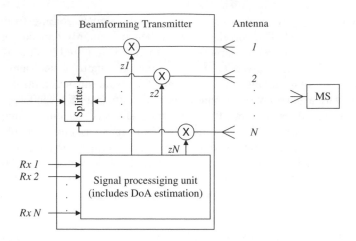

Figure 12.3 Example of a block diagram of a beamforming transmitter with an N-element antenna array

with an N-element antenna array is shown in figure 12.2 (respectively figure 12.3). In the case of a block diagram of a beamforming receiver with an N-element antenna array, a signal processing unit analyses the same signal received from the N antenna elements and computes weights (w_i) that are applied on each path for combining.

On the downlink, the processing is very similar to the uplink. Based on the information measured on the signal received in the uplink, it is possible to estimate the Direction of Arrival (DoA) from the uplink signal and to apply different weights, z_i (amplitude and phase), to the different transmit paths of the same signal, so that the resulting antenna pattern focuses towards the direction of the user.

Since the weights in the downlink depend on the uplink signals, this assumes certain channel reciprocity between the uplink and downlink signals since the BS do not know the downlink spatial channel response. Actually, the reciprocity can more realistically be assumed in the case of the TDD system since the uplink and downlink signals use the same frequency at different time intervals. On the FDD system, the reciprocity is more difficult to assess.

In fact, beamforming technology encompasses several techniques. First implementations of beamforming were based on simple antenna switching mechanisms: in that approach, the elements of the antenna array where simply switched on or off according to the received signals. This has the advantage of simplicity but the possibility for beamforming is limited. Today, beamforming uses an adaptive array: the amplitude and phase of each antenna element can be set independently. This has the advantage of having the possibility to achieve infinity of beams.

With adaptive beamforming, several optimising strategies may be used. The signal processing unit must maximise the received CINR. This can be achieved by having a resulting antenna pattern such that the antenna array creates a null in the direction of arrival of a strong interferer. However, the number of interferers that can be cancelled are limited by the number of elements constituting the array: with N antenna elements, it possible to have at most null $N-1$ interferers. In addition, this technique requires a good knowledge of the radio environment (which may imply additional overheads). This explains why in many implemented systems this method is mainly used in the uplink, where the BS can have maximum knowledge of the radio environment.

Finally, an advanced implementation of beamforming can enable SDMA (Spatial Division Multiple Access). Provided that two or more users are sufficiently separated in space, it is possible to send them at the same time, on the same physical resources, different information on different beams. Nevertheless, the use of SDMA is quite difficult in a mobile environment where MSs that may be well separated at a given moment may be in the same direction at the next moment. More details on smart antennas can be found in [27] and [28].

12.4.1.2 System Design Aspects on BS and MS

All the complexity and intelligence of a beamforming system is inside the base station. In addition to the spatial signal processing unit, the BS includes as many transceivers as the number of antenna elements. Usually, wireless systems implementing beamforming are operating between two to eight antenna elements and transceivers.

On the MS side, the impact for the support of beamforming is minor. Basically, beamforming can be applied to any MS. However, in order to provide better performance, the standards (and in particular IEEE 802.16e) define additional messages/procedures between the BS and the MS. Nevertheless, there is no hardware impact on the terminal side; only some additional software is needed for an optimised beamforming operation.

The beamforming algorithms aim to combine coherently the signal transmitted/received from different antenna elements. In order to achieve this correlation, the antenna elements need to be closely separated. For an optimum performance of beamforming, a spacing of half the wavelength λ is preferred. Consequently, assuming a four-element antenna array at 2.5 GHz would result in an antenna of about 22 to 25 cm width. The array itself is thus of a relatively small size, which is very beneficial for visual impacts on the environment.

Besides, in order to maintain the signal coherence, mechanisms for calibrating the different transmit/receive paths are required, the algorithm for doing this being vendor-specific.

12.4.1.3 Benefits of Beamforming

The benefits of beamforming are manyfold: range increase and power saving at the MS side, interference mitigation and capacity increase.

First, beamforming improves the link budget for the data transmission for both the downlink and the uplink. Indeed, by concentrating the energy in one direction, the resulting antenna gain in one direction is significantly increased (see Figure 12.4). This additional gain is beneficial for improving the coverage of the BS (less sites needed for a deployment) and/or for reducing the power needed by the MS to transmit signals (power saving).

Theoretical gains, compared with a conventional antenna, for an N-element antenna arrays are of $10 \times \log(N)$ for the uplink and $20 \times \log(N)$ for the downlink. For example, with a four-element antenna arrays the gains are respectively of 6 dB (12 dB) for the uplink (respectively the downlink). The gain in the downlink is higher since, on top of the beamforming gain, the power from each transmitter coherently increases. The value of those gains has been validated in many experiments on the field and proves to be in line with the theory [29]. Additional gains are measured in the uplink due to the additional spatial diversity gain.

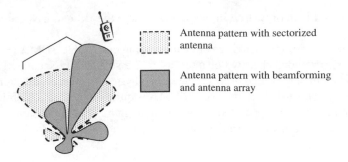

Antenna pattern with sectorized antenna

Antenna pattern with beamforming and antenna array

Figure 12.4   Range extension with beamforming

Second, because the energy is focused in the direction of the user, there is a general inter-ference reduction in a cellular system employing beamforming. Indeed, when beamforming is deployed on the BS of a given geographical area, the beams are oriented as a function of the repartition of the users served in a cell; at one moment, on a given radio resource, a single user is served. As a consequence, the interference created by the communication of this user is only in a restricted angle compared to sectorised antenna deployment (see Figure 12.5).

The angle spread of the main lobe is approximately the total angle of the sector divided by the number of antenna elements N. For example, with a four-element antenna array and a $90°$ antenna, the resulting main lobe width (at $-3\,\mathrm{dB}$) is around $22°$. Therefore, since the users are randomly spread, the beam directions change according to the user locations, which create additional interference diversity gain. The interference reduction is further improved with the use of explicit interference cancellation algorithms.

Two direct consequences of the interference reduction are: a better signal quality and avail-ability across the cell area and a better capacity in the cell for systems using link adaptation. Indeed, since the CINR values are better, the possibility of using a better modulation and coding scheme is higher.

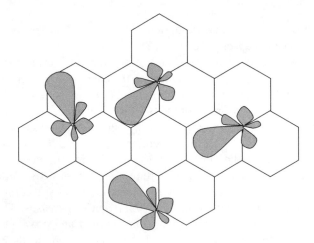

Figure 12.5   Interference reduction with beamforming

12.4.1.4 Support of Beamforming in the IEEE 801.16 Standards

Beamforming is defined in IEEE 802.16-2004 and in 802.16e. This feature is not in the set of the fixed WiMAX profiles. For the mobile WiMAX profiles, this feature is mandatory to be supported by the MS and optional for the BS. For the mobile WiMAX, several mechanisms that enhance the performance and operation of beamforming are provisioned.

In the downlink, in order to be able to beamform several users at the same and on different subchannels, a zone is dedicated (indicated in the DL-MAP). This region, labelled (2) in Figure 12.6, contains permutations with dedicated pilot channels. This means that an MS receiving a burst in this region only takes as valid pilots the pilots associated with the subchannel it has been allocated.

In the uplink, as mentioned previously, the beamforming mechanisms can be applied on any MS. However, to improve the performance and to help the BS in detecting/measuring the interference experienced by the users it wants to serve, a specific signalling zone may be allocated: the uplink sounding zone (indicated by UIUC = 13). Actually, the BS may ask some MS to transmit a signal in this zone so that the BS can evaluate the interference on some subcarriers for those MSs.

Finally, in order to limit the signalling overhead, the WiMAX solution employing beamforming may use the compressed maps and submaps to transmit the common signalling messages (DL-MAP/UL-MAP). Indeed, this permits different modulation, coding and repetition schemes to be applied to several zones in the DL-MAP/UL-MAP message. This solution can also be employed in the case of MIMO or the support of the HARQ.

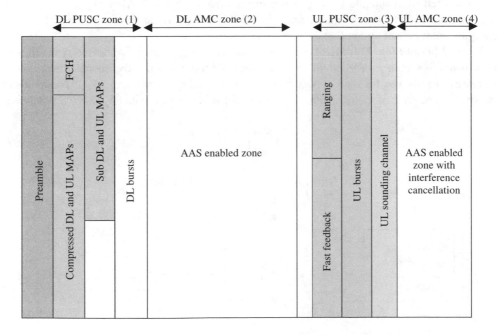

Figure 12.6   Example of a frame with regions supporting AAS operation

12.4.2 MIMO (Multiple-Input Multiple-Output) Solution

12.4.2.1 MIMO Basics

MIMO systems use multiple input and multiple output antennas operating on a single channel (frequency). At the transmitter side, the signal is space–time encoded and transmitted from N_T antennas. At the receiving side, the signals are received from N_R antennas (see Figure 12.7). The space–time decoder combines the signal received by the N_R antennas and transmitted from the N_T antennas after having estimated the channel matrix ($N_T \times N_R$).

The objective of the MIMO solution is to exploit the space and time diversity of the channels on the different radio paths between each combination of transmit/receive antennas to improve the reception sensitivity and/or to improve the channel capacity. There are several families of MIMO solutions. The two extreme ones are the spatial diversity MIMO schemes and the spatial multiplexing MIMO schemes.

Spatial diversity MIMO refers to solutions where the same information is transmitted (after space–time coding) in space and time. The theoretical diversity gain of such a solution is a function of the product of the transmit/receive antennas and is equal to $N_T \times N_R$. For instance, a MIMO system implementing four antennas at the BS side and two antennas at the MS side has a diversity gain of 9 dB. An example of such a scheme is the Alamouti space–time code (STC) [30]. The space–time code of this 2×1 solution is given by the following matrix:

$$A = \begin{bmatrix} S_1 & -S_2^* \\ S_2 & S_1^* \end{bmatrix} \tag{12.6}$$

In matrix A, S_1 and S_2 are the symbols to be transmitted over the air. At symbol k, symbol S_1 is transmitted from antenna 1 and S_2 is transmitted from antenna 2. At symbol $k + 1$, symbol $-S_2^*$ is transmitted from antenna 1 and symbol S_1^* is transmitted from antenna 2 (S_i^* is the complex conjugate of symbol S_i).

Spatial Multiplexing (SM) MIMO refers to solutions where, during a symbol interval, different information is sent in parallel on different antennas. With this scheme, theoretically the capacity increases linearly as a function of N, N being the minimum between N_T and N_R. An example of the space–time code of such a solution is given by the following formula for a 2×2 SM scheme:

$$B = \begin{bmatrix} S_1 \\ S_2 \end{bmatrix} \tag{12.7}$$

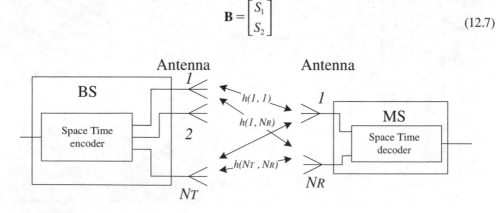

Figure 12.7   Generic MIMO block diagram for the downlink

With the SM code, the capacity of the channel and the burst transmitted rate are increased. However, this only happens under very good CINR conditions and a highly uncorrelated channel.

In addition, several MIMO schemes exist that are a mix between SM and spatial diversity schemes. The diversity order and capacity increase depends on the space–time code and number of antennas.

More recently, MIMO schemes using pre-coding have been defined. In these cases, the space–time code depends on a feedback from the receiver on the channel states. Indeed, this solution requires a closed-loop operation and additional signalling between the receiver and the transmitter.

Finally, MIMO can also be generated from signals transmitted from different BSs (virtual MIMO). This requires time synchronisation of the BS but also a synchronisation of the scheduler of the BS involved in the transmission. More details on the MIMO scheme can be found in Reference [31] or [32].

12.4.2.2 System Design Aspects of BS and MS

As indicated in Figure 12.7, MIMO operation has a significant impact on the design of the BS and the MS. Indeed, in addition to the MIMO codes, the algorithms for encoding and decoding MIMO signals, there is the requirement to implement several transmitter and several receivers both at the BS and MS sides. This may be critical for the MS side. Indeed, implementing several receiver chains implies several antennas on a small device and additional hardware and processing power. Moreover, several transmitters at the MS side also mean significant additional power consumption. Efficient implementation of MIMO at the MS side has then some technological challenges to be solved.

The preferred antenna configuration for MIMO is when the multiple antennas are perfectly uncorrelated. In that case, the performances of MIMO are optimal. Good correlation is obtained if at the BS side the antennas are separated by 10 to 20 λ and at the MS side by at least λ. The latter requirement makes solutions with more than two antennas at the MS impractical.

In some cases, the additional effort needed for the space–time decoder is significant. This also imposes some limits on the number of transmit antennas at the BS side.

12.4.2.3 Benefits of MIMO

Depending on the scheme, the benefits of MIMO can be to improve the receiver sensitivity and/or to multiply the capacity and the peak rates. However, as stated previously, to get the best performance from MIMO, the different receive/transmit channels should be highly uncorrelated. This is achieved in the environment with very high scattering and signals coming from a very large angular spread (e.g. in a microcellular environment).

Especially, the SM schemes are very sensitive to correlation since in the case of a correlated channel, the SM scheme creates additional interference.

12.4.2.4 Support of MIMO in the IEEE 801.16 Standards

The IEEE 802.16-2004 standard provides a few supports for MIMO. Only the Alamouti scheme as defined by Equation (12.6) is defined and is not mandatory in the profile for fixed/nomadic WiMAX systems. On the contrary, IEEE 802.16e provides extensive support for MIMO.

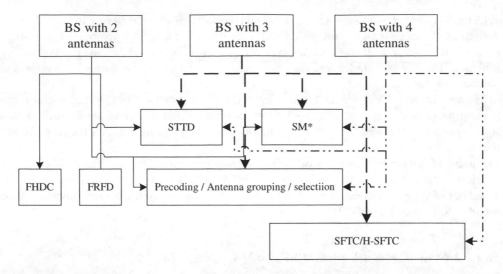

Figure 12.8 Downlink MIMO support in IEEE 802.16e (based on Reference [2]). STTD = Space Time Transmit Diversity, SM = Spatial Multiplexing (*SM mode requires at least two antennas at the MS side), FHDC = Frequency Hopped Diversity Coding, FRFD = Full Rate Full Diversity, SFTC = Space Frequency Time Code and H-SFTC = Hybrid SFTC

Out of the schemes defined in the standard, the WiMAX Forum has selected two MIMO schemes for the downlink with two antennas at the BS side and two antennas at the MS side. The schemes are those defined in equations (12.6) and (12.7). Downlink MIMO support in IEEE 802.16e is shown in figure 12.8. In the downlink, the mobile WiMAX solution can either benefit from the additional diversity gain or doubled capacity and peak rate.

In the uplink, among all the options included in the standard (see Figure 12.9), only one scheme has been selected for mobile WiMAX: collaborative SM. With collaborative SM, two MSs, equipped with a single transmitter, send different data information at the same time on the same channels. In such a mode, the users can be discriminated with the help of pilot

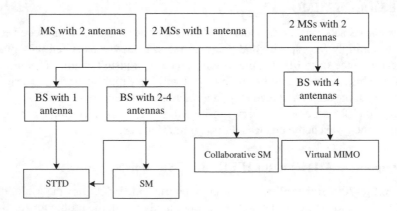

Figure 12.9   Uplink MIMO support in IEEE 802.16e-2005. (Based on Reference [2].)

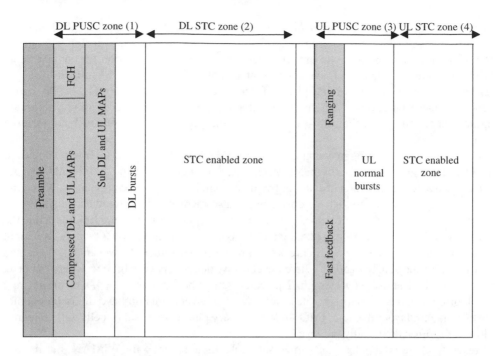

Figure 12.10 Example of a frame with regions supporting the MIMO operation

symbols, which use different tones for each user (each MS transmits half of the total available pilots).

In the uplink, MIMO can improve the capacity of the mobile WiMAX system by providing sufficient signal quality. Of course, since the use of SM is sensitive to the radio condition, a MIMO solution may include the solution to switch to the best MIMO mode depending on the signal quality.

Indeed, MIMO operation implies that the signals transmitted over the air have a specific format. In order to mix users operating with MIMO with other users in one frame, MIMO zones are defined (indicated in the DL-MAP message). For instance, Figure 12.10 is an example of a frame mixing MIMO and non-MIMO zones. In this figure, zone (1) includes the common signalling channels and some downlink bursts (using a PUSC permutation). Zone (2) is the MIMO zone for the downlink. Zone (3) is the zone in the uplink for users who cannot use collaborative MIMO (e.g. because their estimated signal-to-interference level is too weak). Zone (3) also includes common uplink signalling messages for ranging, fast feedback, etc. Finally, zone (4) is the uplink MIMO zone.

12.4.3 About the Implementation of Advanced Antenna Technologies

As discussed in previous sections, beamforming and MIMO are key solutions to provide spectral efficiency for a mobile WiMAX system, making it superior to the spectral efficiency of other existing cellular systems. Also, both MIMO and AAS technologies (in particular in the light of the choices made by the WiMAX Forum) have different and complementary benefits.

Beamforming (or AAS), since it provides link budget gains for both the uplink and the downlink, is well suited for environments that are coverage-limited. For instance, comparing a system implementing a two-branch receive diversity with a solution for beamforming with four antenna elements can save between 40 and 50 % of radio sites. Beamforming could then be recommended for the first phase of deployment of a mobile WiMAX. This corresponds to macrocell coverage where the signal at the BS side is received with really limited angular spread (no more than 20–30°), which is the preferred situation for beamforming algorithms.

Also, using beamforming may enable mobile WiMAX solutions at 3.5 GHz to be deployed since the extra gains provided by beamforming can compensate for the additional propagation losses due to transmission at a higher frequency band. As discussed previously, deploying beamforming not only increases the coverage but also increases the offered capacity.

The mobile WiMAX solution does not provide a diversity gain in the uplink (which is the weak link in WiMAX, as in most cellular systems). Consequently, MIMO solutions cannot be deployed to improve the range in areas that are coverage-limited. However, MIMO can be deployed advantageously in areas where the capacity demand is very high or where the peak rate offer to the end user is very high. This can occur in hot-spot areas (e.g. a business area) or indoor environments. Actually, these very dense environments do have the radio characteristics that are favourable to MIMO. Indeed, in very dense areas, small cells with antennas below the rooftop are usually deployed.

Presently, there is no single and universal deployment strategy for WiMAX, but the use of antenna technology (either beamforming or MIMO or both) does provide significant differentiation with other existing wireless systems.

12.5 Multicast Broadcast Services (MBS)

Multicast Broadcast Services (MBS) may be required when multiple MSs connected to a BS receive the same information or when multiple BSs transmit the same information. Indeed, this allows resource to be saved by allocating a single radio pipe for all users registered to the same service instead of allocating as many pipes as there are users. This is of particular interest for the broadcast TV type of application where, at the same time, several users under the same coverage area are connected to the same service (in this case, a TV channel).

12.5.1 Multi-BS Access MBS

The mobile WiMAX system supports MBS as an optional feature for the BS. When the MBS feature is supported, the Multi-BSs Access mode is implemented, as defined in the IEEE 802.16e standard [2].

In a Multi-BS MBS system, several BSs in the same geographical area transmit the same broadcast/multicast messages at the same time on the same frequency channel. These BSs actually belong to MBS_ZONE. An MBS_ZONE is a unique identifier, which is transmitted from each BS of the set on the DCD message. An example of an MBS_ZONE allocation is provided in Figure 12.11. It has to be noted that a BS may belong to different MBS_ZONEs.

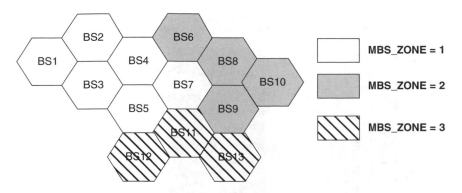

Figure 12.11 Example of MBS zone deployment and MBS_ZONE allocations. All the MS registered to an MBS_ZONE can receive MBS signals from any BS of the MBS_ZONE

A Multi-BS MBS operation requires from the BSs belonging to the same MBS_ZONE:

- time synchronisation (frame number and symbol level);
- use of the same CID for broadcast/multicast messages;
- use of the same Security Association (SA) for the encryption of the broadcast/multicast messages.

Using a Multi-BS MBS solution provides two major advantages. First, the MSs that successfully registered to a MBS service can receive MBS information from any BS of the MBS_ZONE without needing to register to a specific BS of that zone (even MS previously in the Idle mode). Secondly, the MS receives MBS signals from multiple BSs simultaneously. This provides a macrodiversity gain and performance enhancement for the MBS signals. This is actually similar to the Single Frequency Network (SFN) concept that can be found in DVB-H systems.

12.5.2 MBS Frame

When the MBS is activated on a BS, it can use a dedicated frequency or it can use only a dedicated zone in the frame, as illustrated in Figure 12.12. The descriptions of existing MBS zones in a frame are indicated using the DL-MAP in the extended DIUC 2 field (DUIC = 14) by MBS_MAP_IE, which includes (among others) the MBS_ZONE identifier, the symbol offset where the MBS zone starts, the permutation to be used (PUSC or other) and the size of the MBS zone (in symbols and subchannels).

Each MBS zone starts with MBS_MAP, which describes the MBS connections available in the MBS zone. In particular, it contains, among others, the CIDs used by the multicast/broadcast connections, the channel configurations (modulation, coding and eventually repetition coding) and the logical channel ID. Logical channel IDs are used to distinguish the different MBS connections inside a MBS zone that belong to the same MBS CID (for instance, this may be used to differentiate between different TV channels with different subscription rights).

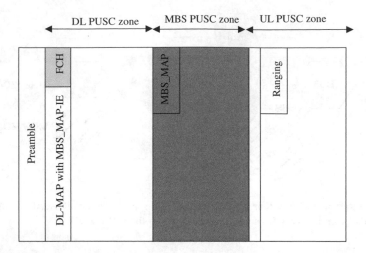

Figure 12.12   Example of the 802.16 frame with the MBS service zone. The presence of the MBS zone is indicated in the DL-MAP message (in a MBS_MAP_IE field). The exact details of the MBS zone are then described in the MBS_MAP message at the beginning of the MBS zone

The MBS PDUs in the MBS region follow the order of the combination of multicast CID/ logical channel CID, as described by the service association between the MS and the BS. Of course, from the time when the MS goes into the Idle mode or when the MS moves across different BS areas of the same MBS_ZONE there is no need for the MS to be registered to the BS to receive the MBS content. The mapping between the multicast CID and the logical channel CID must be the same in the cells belonging to the same MBS_ZONE. The procedure to provide consistency of the identities used by the MBS channels is beyond the scope of the IEEE 802.16 standard.

13

WiMAX Architecture

With the contribution of Jérôme Brouet, Alcatel, France

13.1 The Need for a Standardised WiMAX Architecture

13.1.1 Supporting Working Groups and Documents

The IEEE 802.16 standard only defines the PHY and MAC layers. Consequently, in order to ensure intervendor internetwork interoperability, for operations such as roaming, it is important to define standards over the widest range of interfaces and equipments, as in 3GPP and 3GPP2 mobile standards. Then, in addition to the alignment of the radio access features for multivendor interworking between base stations and terminals based on the 802.16 standards [1,2], the WiMAX Forum charter also aims to deliver a framework for a high-performance end-to-end IP network architecture to support fixed, nomadic, portable and mobile users (see Table 13.1 for service type definitions).

The WiMAX architecture is based on the use of standardised IP protocols and is compatible with service frameworks such as the IP Multimedia Subsystem (IMS). Two working groups from the WiMAX Forum organisation (see Chapter 2) define the architecture and associated functionalities: the Network Working Group (NWG) creates the network specifications and the Service Provider Working Group (SPWG) helps to define requirements and priorities.

The set of specifications issued by these two groups includes several alternatives for mapping different required functionalities to physical equipments, allowing at the same time added value manufacturer-dependent implementation choices and also interoperability points at the network level, thanks to standardised open interfaces (or reference points in the WiMAX wording) [21, 10, 34]. The latest NWG drafts are publicly accessible at http://www.wimaxforum.org/technology/documents.

The specifications are organised by release (currently, Release 1 will be finalised by Q3'06) and consists of a set of three-stage documents:

- Stage 1 document defines the requirements for the WiMAX architecture.
- Stage 2 documents describe the network reference model, the reference points and also include some informative parts for interworking between a WiMAX network and another network (e.g. a DSL network) [21]. The WiMAX network architecture is evidently also applicable to standalone deployments (not just for interworking scenarios).

Table 13.1   WiMAX service definition. (Based on Reference [33].)

Service type	Service area	Speed	Handover	Comment
Fixed	Single BS	Stationary	No; yet may involve BS reselection due to link conditions	Usually, fixed outdoor terminals
Nomadicity	Any BS	Stationary	No; yet may involve BS reselection due to link conditions	Indoor/Outdoor use; usually self-install
Portability	Any BS	Mobile	Yes	Session continuity for non-real-time applications
Simple mobility	Any BS	Mobile	Yes	Performance for non-real-time applications and some support of real-time applications
Full mobility	Any BS	Mobile	Yes	Session continuity for real-time applications

- Stage 3 documents define the details of the protocols and other procedures to be implemented in a WiMAX end-to-end network [34].

13.1.2 High-level Architecture Requirements

The WiMAX reference architecture has been created having in mind various types of requirements:

- a high-performance packet-based network with functional split, ensuring maximum flexibility based on standard protocols from IEEE and IETF;
- the support of a full scale of services and applications;
- the support of roaming and interworking with other fixed/mobile networks.

In terms of services and applications, a WiMAX network is designed to be able to support:

- voice (using VoIP), multimedia (using IMS) and other mandatory regulatory services such as emergency calls;
- access to a large variety of application service providers;
- interfacing with a variety of interworking and media gateways for translating legacy services (e.g. circuit voice, MMS) to IP and to transport them over WiMAX radio access networks;
- delivery of IP MBS, multicast and broadcast services (see Chapter 12).

In addition, considering network interworking and roaming, several deployment scenarios must be supported:

- loose coupling with existing wired (e.g. DSL) or wireless networks (e.g. 3GPP or 3GPP2 mobile networks);
- global roaming between WiMAX operators (this includes among others a consistent use of the AAA, Authentication Authorization and Accounting, among WiMAX operators for authentication and billing);
- a variety of user authentication methods (username/password, digital certificates, SIM based).

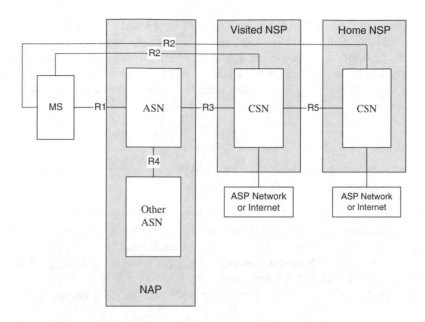

Figure 13.1 WiMAX network reference model with components (MS/ASN/CSN), reference points (R1 to R5) and actors (NAP/NSP/ASP) [10]

The resulting reference architecture and functional split is developed further in the next sections of this chapter.

13.2 Network Reference Model

13.2.1 Overview and Definitions

The WiMAX network reference model (see Figure 13.1) consists of three components interconnected by standardised interfaces or reference points R1 to R5. The three components are:

- MS (Mobile Station);
- ASN (Access Service Network);
- CSN (Connectivity Service Network).

The Mobile Station (MS) is a generic mobile equipment providing connectivity between a subscriber equipment and a WiMAX BS. The Access Service Network (ASN) includes the set of functionalities that provides radio access connection to WiMAX subscribers. One or several ASN, interconnected through reference point R4, may be deployed by a Network Access Provider (NAP). A NAP provides radio access infrastructure to one or several Network Service Providers (NSP). The NSP is a business entity that enables IP connectivity and WiMAX services to WiMAX subscribers in accordance with established Service Level Agreements (SLA). The NSP deploys the Connectivity Service Network (CSN), which provides the IP connectivity for WiMAX subscribers. On the radio access network side, the WiMAX services are provided through contractual agreements with one or several NAPs.

On the application side, WiMAX services are delivered thanks to contractual agreements with Application Service Providers (ASP) and/or through direct connection to the Internet. Additionally, an NSP in a given country may have roaming agreements with many other NSPs, which may be in other countries. Hence, a WiMAX subscriber may be attached to a Home NSP (H-NSP) or to a Visited NSP (V-NSP), a NSP with whom its home NSP has a roaming agreement.

13.2.2 ASN Reference Model and Profiles

The ASN includes all the functionalities that enable radio connectivity to WiMAX subscribers. As a consequence, the ASN mainly provides:

- Layer 2 connectivity to the WiMAX subscribers (through the WiMAX air interface);
- Radio Resource Management (RRM) mechanisms such as handover control and execution;
- paging and location management (in the case of portable/mobile services);
- relay functions to the CSN for establishing Layer 3 connectivity with WiMAX subscribers (e.g. IP address allocation, AAA procedures);
- tunnelling of data and signalling between the ASN and the CSN though reference point R3;
- network discovery and selection of the preferred NAP/NSP.

The ASN usually consists of several base stations (BS) connected to several ASN gateways (ASN-GW), as depicted in Figure 13.2. Inside the ASN, two additional reference points are described in WiMAX architecture specification Release 1, but only in an informative manner: R6 between BS and ASN-GW and R8 between different BSs [21]. Those interfaces

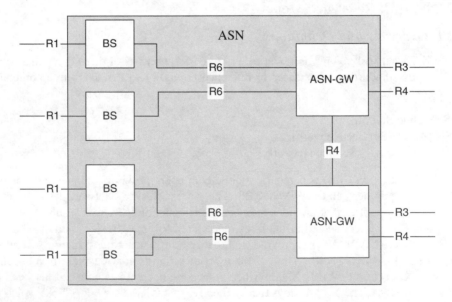

Figure 13.2 Generic ASN reference model. R3 is between the ASN and CSN while the external R4 is between ASNs

should be the basis for further interoperability points between BS and ASN-GW of WiMAX vendors in subsequent architecture releases.

13.2.2.1 Base Station (BS)

The BS is the entity that implements PHY and MAC features as defined in the IEEE 802.16 standards. As defined by the WiMAX Forum technical groups, the BS is also in charge of the scheduling of user and signalling messages exchanged with the ASN-GW through the R6 interface (see below for R6). It may also incorporate other access functions according to the ASN profile (see Table 13.2).

In a WiMAX access network, a BS instance is defined by a sector and a frequency assignment. In the case of a multiple frequency assignment in a sector, the sector includes as many BS instances as frequency assigned. This is similar to 3GPP UMTS or 3GPP2 networks.

The connectivity to multiple ASN-GW may be required in the case of load balancing or for redundancy purposes.

13.2.2.2 ASN Gateway (ASN-GW)

The ASN-GW is a logical entity that includes:

- control function entities paired with a corresponding function in the ASN (e.g. in a BS instance), a resident function in the CSN or a function in another ASN;
- bearer plane routing or a bridging function.

The ASN-GW acts as a decision point for nonbearer plane functions (example: radio resource management) and as an enforcement point for bearer plane functions. For implementation purposes, decomposition of ASN functions into these two groups is optional. If this decomposition is done, the two groups are separated by reference point R7. As in any telecom system, the ASN-GW may be designed to provide redundancy and load balancing between different ASN-GW boxes.

13.2.2.3 ASN Profiles

There are three choices for ASN implementations: ASN profiles A, B and C (see Figure 13.3). According to the profiles, some functionalities are implemented by the BS or the ASN-GW (profiles A and C, see Table 13.2) or by any box in the case of a profile B ASN. In particular, profile A implementation includes:

- handover control inside the ASN-GW, which permits easier control of the radio resources during the handover procedure and preparation of the Layer 3 path re-route;
- Radio Resource Control (RRC) inside the ASN-GW (load balancing is possible between BSs).

ASN anchored mobility points among BSs are achieved through R4 and R6 reference points. This mobility refers to handovers where the anchor point for the MS in the ASN does not change (see Chapter 14 for handovers in the IEEE 802.16 standard).

Profile C implementation has the same structure as profile A, with the major difference being that BSs have much more functionalities. A profile C implementation includes:

Table 13.2 Split of ASN functions between BS and ASN-GW for profile A and profile C ASN. (Based on Reference [21].)

Function category	Function	Exposed protocols, primitives	Associated reference point	Profile A, ASN entities	Profile C, ASN entities
Security	Authenticator	Authentication relay primitives	R6	ASN-GW	ASN-GW
	Authenticator relay	Authentication relay primitives	R6	BS	BS
	Key distributor	AK[a] transfer primitives SA-TEK[b] handshake primitives	R6	ASN-GW	ASN-GW
	Key receiver	AK[a] transfer primitives SA-TEK[b] handshake primitives	R6	BS	BS
Intra-ASN mobility	Data path function (type 1 or 2)	Data path control primitives	R6	ASN-GW and BS	ASN-GW and BS
	Handover functions	Handover control primitives	R6	ASN-GW and BS	ASN-GW and BS
	Context server and client		R6	ASN-GW and BS	ASN-GW and BS
Layer 3 mobility	MIP FA[c]	Client MIP	R6		ASN-GW
Radio resource management	RRC[d]	RRM primitives	R6	ASN-GW	BS
	RRA[e]	RRM primitives	R6	BS	BS
Paging	Paging agent	Paging and Idle mode primitives	R6	BS	BS
	Paging controller	Paging and Idle mode primitives	R6	ASN-GW	ASN-GW
QoS	SFA[f]	QoS primitives	R6	ASN-GW	ASN-GW
	SFM[g]			BS	BS

[a] AK = Authentication Key (transferred during initial network entry).

[b] SA-TEK = Security Association-Traffic Encryption Key (during a three-way handshake procedure between the MS and BS at initial network entry).

[c] MIP FA = Mobile IP Foreign Agent (used for inter-ASN-GW mobility).

[d] RRC = Radio Resource Controller (entity that takes decisions on modifications of radio resources based on radio measurements, or any other information).

[e] RRA = Radio Resource Agent (entity located in a BS that maintains a database of radio resource indicators).

[f] SFA = Service Flow Authorisation (see the QoS section).

[g] SFM = Service Flow Management (see the QoS section).

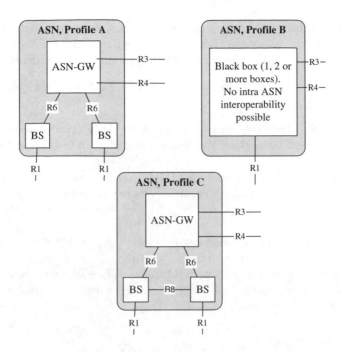

Figure 13.3   The ASN profiles A, B and C

- handover control inside the BS;
- Radio Resource Control (RRC) inside the BS;
- ASN anchored mobility points among the BS achieved through R4 and R6 reference points, as in profile A.

Profile B does not specify any functional split. The split of functions between boxes, if any, is left to vendor choice. As a consequence, no intra-ASN interoperability is possible with profile B.

13.2.3 CSN Reference Model

The Connectivity Service Network (CSN) consists of all the functions/equipment that enable IP connectivity to WiMAX subscribers. As a consequence, the CSN includes the following functions:

- user connection authorisation and Layer 3 access (IP address allocation for user sessions, AAA proxy server and functions);
- QoS management (policy and admission control based on user profiles);
- mobility support based on Mobile IP (HA (Home Agent), function for inter-ASN mobility);
- tunnelling (based on IP protocols) with other equipment/networks (ASN-CSN tunnelling support, inter-CSN tunnelling for roaming);
- WiMAX subscriber billing;

- WiMAX services (Internet access, location-based services, connectivity for peer-to-peer services, provisioning, authorisation and/or connectivity to IMS, facilities for lawful intercept services such as CALEA (Communications Assistance Law Enforcement Act)).

To accomplish those functions, the CSN, deployed by the NSP, may include the following equipment:

- routers (with eventually an HA function for the inter-ASN gateway mobility);
- DNS (Domain Name System)/DHCP servers for IP address resolution and user IP address configurations;
- AAA proxy/servers and user database for WiMAX user access authentication/authorisation/accounting and provisioning, with some reports mentioning the use of a (central) Home Location Register (HLR) as in some other cellular systems (e.g. GSM, UMTS) in order to implement AAA functions;
- interworking gateways for integration/interoperability of a WiMAX network with another network (e.g. a 3GPP wireless network or a PSTN);
- firewalls for providing protection to the WiMAX network equipments by enforcing access and filter policies on the traffic to and from an external network (especially used for denial of services detection/prevention).

13.2.4 Reference Points

The WiMAX network reference model defines several reference points (RPs) between several entities in the WiMAX network (see Figures 13.1 and 13.3). Those RPs introduce interoperability points between equipments from different vendors. In the scope of the Release 1 WiMAX network architecture, there are six mandatory RPs (R1 to R6) and two informative RPs (R7 and R8).

13.2.4.1 Reference Point R1 (Normative)

Reference point R1 refers to the radio interface between the MS and the ASN. It consequently includes all the physical and MAC features retained in WiMAX profiles from the IEEE 802.16 standard. R1 carries both user traffic and user control plane messages.

13.2.4.2 Reference Point R2 (Normative)

Reference point R2 is a logical interface between the MS and the CSN. It contains all the protocols and other procedures that are involved with authentication (user and device), service authorisation and IP host configuration management. This logical interface is established between the MS and the H-NSP and some protocols (such as the IP host address management) may be performed by the visited NSP in the case of roaming.

13.2.4.3 Reference Point R3 (Normative)

Reference point R3 is the logical interface between the ASN and the CSN. It conveys both control plane messages, (AAA methods, policy enforcement methods for end-to-end QoS, mobility management messages e.g. for MS relocation) and data plane information through tunnelling between the ASN and the CSN.

13.2.4.4 Reference Point R4 (Normative)

Reference point R4 interconnects two ASNs (ASN profile B) or two ASN-GWs (ASN profiles A or C). It conveys both control and data plane messages, especially during the handover of a WiMAX user between ASNs/ASN-GWs or during location update procedures in the Idle mode. This RP is presently the only interoperable point between ASNs from different vendors.

The tunnelling method recommended is to use IP in IP tunnelling mode for R4 based on the GRE (Generic Routing Encapsulation) protocol [35].

13.2.4.5 Reference Point R5 (Normative)

Reference point R5 is the interface interconnecting two CSNs. It consists of the set of control and data plane methods between the CSN in the visited NSP and in the home NSP.

13.2.4.6 Reference Point R6 (Normative for Profiles A and C)

Reference point R6 is normative in WiMAX architecture Release 1 in the context of specific ASN profiles. In Release 1, ASN profiles A and C decompose the ASN into BS and ASN GW. In each case, normative procedures over R6 are specified in Release 1. R6, of course, is not applicable to profile B.

Reference point R6 connects the BS and the ASN-GW. It conveys both control messages (for data path establishment, modification, control, release in accordance with MS mobility) and data plane (intra-ASN data path between BS and ASN-GW) information. The tunnelling method used is GRE, MPLS or VLAN (other tunnelling methods may be proposed in a later phase as well). This interface can also convey, in combination with R4, the MAC state information carried by the R8 reference point when R8 interoperability between BSs is not available.

13.2.4.7 Other Informative Reference Points

In the ASN, there are currently two additional interfaces (R7 and R8) defined for further interoperability points in the scope of the next WiMAX architecture releases. Those interfaces are only informative in the frame of WiMAX architecture Release 1.

Reference point R7 is an optional logical interface between the decision function and the enforcement function in the ASN-GW.

Reference point R8 is a logical interface between BSs. It conveys the control plane flow exchange that is used for enabling a fast and efficient handover between BSs. Optionally, R8 may also convey data plane information during the handover phase. It should be noted that a direct physical interface is not actually required between BSs. The R8 methods can indeed be conveyed through, for example, the ASN-GW.

13.3 Network Functionalities

This section describes the main functions achieved by the WiMAX end-to-end system.

13.3.1 Network Discovery and Selection

This function is required for nomadic, portable and mobile WiMAX services (see Table 13.1) where in the same geographical area the MS may have radio coverage access to an ASN managed by a single NAP and shared by several NSPs or coverage access to several ASNs managed by several NAP/NSPs. To perform network discovery and selection, the MS (which has been pre-provisioned with configuration information at subscription) performs a four-step process:

- NAP discovery. During the scanning of the DCD of the different BSs in the coverage reach from the MS, the MS detects the 'operator ID' in the BSID field.
- NSP discovery. The MS discovers the available NSP through the list NSP ID which is broadcast by the ASN as part of the system information identity message. NSP discovery is also possible via solicited request/response messages.
- NSP enumeration and selection. Based on the dynamic information obtained in the coverage area and the configuration information from the subscription, the MS selects the appropriate NSP. Manual configuration of the NSP may also be available in the case of a visited NSP.
- ASN attachment. After selection of the NSP and associated ASN, the MS indicates its NSP selection by sending an NAI (Network Access Identifier) message used by the ASN to determine the next hop AAA where an MS AAA packet should be routed to.

13.3.2 IP Addressing

WiMAX networks support IPv4 and IPv6 addressing mechanisms. At the end of the procedure, a PoA (Point of Attachment) IP address is delivered to the MS. The IP allocation address modes depend on the WiMAX access service types.

In the case of IPv4, the dynamic PoA configuration is based on DHCP [17, 18]. The DHCP proxy may reside in the ASN and the DHCP server in the CSN. In the case of IPv6, stateful IP address allocation is based on DHCPv6 [36]. The DHCP server resides in the CSN and the DHCP proxy may reside in the ASN. For the stateless CoA (Care of Address), IP address allocations RFC 2462 [37] and RFC 3041 [38] are used.

PoA IP address methods according to the WiMAX access services and IP version are summarized in Table 13.3.

13.3.3 AAA Framework

The AAA framework follows the IETF specifications and includes the following services:

- authentication: device, user or combined user/device authentication;
- authorisation: user profile information delivery for sessions, mobility and QoS;
- accounting: delivery of information for pre-paid/post-paid services.

Table 13.3 PoA IP address methods according to the WiMAX access services and IP version

Service type	PoA IP address scheme (IPv4)	PoA IP address scheme (IPv6)
Fixed access	Static or dynamic	Static or stateful autoconfiguration
Nomadic access	Dynamic	Stateful or stateless autoconfiguration
Mobile access	DHCP for P-MIP[a] terminals MIP based for C-MIP[b] terminals	Stateful or stateless autoconfiguration

[a] P-MIP=Proxy-Mobile IP mode.
[b] C-MIP=Client-Mobile IP mode.

Authentication and authorisation procedures are based on the EAP (Extensible Authentication Protocol) [39]. Between the MS and the ASN (the authenticator function), EAP runs over PKMv2, which enables both user and device authentication. Between the AAA server and the ASN, the EAP runs over RADIUS [40].

Accounting is also based on RADIUS messages. WiMAX Release 1 offers both offline (post-paid) and online (pre-paid) accounting. In the case of offline accounting, the ASN generates UDRs (Usage Data Records), which are a combination of radio-specific parameters and IP-specific parameters. The serving ASN then sends the UDR to the RADIUS server.

13.3.4 Mobility

The mobility procedures are divided into two mobility levels:

- ASN anchored mobility procedures. This refers to MS mobility where no CoA address update is needed, also known as micromobility.
- CSN anchored mobility procedures. The macromobility between the ASN and CSN is based on mobile IP protocols running across the R3 interface.

CSN anchored mobility implies that, in the case of IPv4, the MS changes to a new anchor FA (Foreign Agent). WiMAX systems must support at least one of the following mobile IP schemes:

- Proxy-MIP. In this case, the MS is unaware of CSN mobility management activities and there is no additional signalling/overhead over the air to complete the CSN mobility.
- Client MIP (CMIPv4). In this case, the MIP client in the MS participates in inter-ASN mobility.

13.3.5 End-to-End Quality of Service

The IEEE standard defines the QoS framework for the air interface (see Chapter 11). The WiMAX architecture specifications extend the QoS framework to the complete network where many alternatives for enforcing the QoS on Layer 2 or Layer 3 may exist.

The end-to-end QoS framework relies on functions implemented in the CSN (PF (Policy Function) and AF (Application Function)) and in the ASN (SFM (Service Flow Management) and SFA (Service Flow Authorisation)). In the CSN, the AF triggers a service flow trigger to the PF based on the information sent by the MS with whom it communicates. The PF then evaluates service requests against a policy database in the NSP. In the ASN, the SFA communicates with the PF and is responsible for evaluating the service request against user QoS profiles. The SFM (located in the BS and responsible for creation, admission, modification and release of 802.16 service flows) mainly consists of an admission control function, which decides, based on available radio resource and other local information, whether a radio link can be created.

14

Mobility, Handover and Power-Save Modes

14.1 Handover Considerations

One of the major goals of the 802.16e amendment is to introduce mobility in WiMAX. Consequently, mobile WiMAX profiles are based on 802.16e. Mobility is based on handover. Handover operation (sometimes also known as 'handoff') is the fact that a mobile user goes from one cell to another without interruption of the ongoing session (whether a phone call, data session or other). The handover can be due to mobile subscriber moves, to radio channel condition changes or to cell capacity considerations. Handover is a mandatory feature of a cellular network. In this chapter the handover (HO) is described as defined in 802.16e.

In 802.16e, the two known generic types of handover are defined:

- Hard handover, also known as break-before-make. The subscriber mobile station (MS) stops its radio link with the first BS before establishing its radio link with the new BS. This is a rather simple handover.
- Soft handover, also known as make-before-break. The MS establishes its radio link with a new BS before stopping its radio link with the first BS. The MS may have two or more links with two or more BSs, which gives the soft handover state. The soft handover is evidently faster than the hard handover.

Two types of soft handover are then defined in 802.16e [2]:

- Fast BS Switching (FBSS). This is a state where the MS may rapidly switch from one BS to another. The switch is fast because the MS makes it without realising the complete network entry procedure with regard to the new BS.
- Macro Diversity HandOver (MDHO). Transmissions are between the MS and more than one BS.

In the mobile WiMAX profiles, only the hard handover is mandatory. The FBSS and MDHO are optional. The 802.16 standard also indicates that the support of the MDHO or FBSS is optional for both the MS and the BS.

WiMAX: Technology for Broadband Wireless Access   Loutfi Nuaymi
© 2007 John Wiley & Sons, Ltd

Handover has challenging objectives (see Section 13.1 for handover requirements as a function of the WiMAX access type). First, it has to be fast enough, of the order of 50 ms or 150 ms. There are also security requirements, as some attacks are possible at the occasion of the handover procedure. Finally, the handover does not have only Layer 2 considerations. Layer 3 considerations are also needed, as mentioned in Chapter 13. Hence, the handover is not independent of the architecture.

14.2 Network Topology Acquisition

14.2.1 Network Topology Advertisement

A BS broadcasts information about the network topology using the MOB_NBR-ADV (Neighbour ADVertisement) MAC management message [2]. This message provides channel information about neighbouring BSs normally provided by each BS's own DCD/UCD message transmissions. The MOB_NBR-ADV does not contain all the information of neighbouring BSs, UCD and DCD. The standard indicates that a BS may obtain that information over the backbone and that availability of this information facilitates MS synchronisation with neighbouring BS by removing the need to monitor transmission from the neighbouring (handover target) BS for DCD/UCD broadcasts. The BSs will keep mapping tables of neighbour BS MAC addresses and neighbour BS indexes transmitted through the MOB_NBR-ADV message, for each configuration change count, which has the same function as for the DCD message.

BSs supporting mobile functionality must be capable of transmitting a MOB_NBR-ADV MAC management message at a periodic interval to identify the network and define the characteristics of the neighbour BS to a potential MS seeking initial network entry or handover. The standard indicates that the maximum value of this period is 30 seconds.

14.2.2 MS Scanning of Neighbour BSs

A scanning interval is defined as the time during which the MS scans for an available BS [2]. A BS may allocate time intervals to the MS for the purpose of MS seeking and monitoring suitability of neighbour BSs as targets for a handover. MS scanning of neighbour BSs is based on the following MAC Management messages: MOB_SCN-REQ, SCaNning interval allocation REQuest, MOB_SCN-RSP, SCaNning interval allocation Response, MOB_SCN-REP and SCaNning result REPort.

The MOB_SCN-REQ message is sent by the MS to request a scanning interval for the purpose of seeking available BSs and determining their suitability as targets for HO. In the MOB_SCN-REQ message the MS indicates a group of neighbour BSs for which only Scanning or Scanning with Association are requested by the MS. The *Neighbour_BS_Index* of the MOB_SCN-REQ message corresponds to the position of BSs in the MOB_NBR-ADV message. In this message, the MS may also request the scanning allocation to perform scanning or noncontention Association ranging. Association is an optional initial ranging procedure occurring during the scanning interval with respect to one of the neighbour BSs (see the following section).

Upon reception of the MOB_SCN-REQ message, the BS responds with a MOB_SCN-RSP message. The MOB_SCN-RSP message can also be unsolicited. The MOB_SCN-RSP message either grants the requesting MS a scanning interval that is at least as long as that requested by the MS or denies the request. In the MOB_SCN-RSP message the BS indicates a group of neighbour BSs for which only Scanning or Scanning with Association are recommended by the BS.

Following reception of a MOB_SCN-RSP message granting the request, an MS may scan for one or more BSs during the time interval allocated in the message. When a BS is identified through scanning, the MS may attempt to synchronise with its downlink transmissions and estimate the quality of the PHY channel.

The BS may negotiate over the backbone with a BS Recommended for Association (in the MOB_SCN-REQ message) the allocation of unicast ranging opportunities. Then the MS will be informed on *Rendez vous time* to conduct Association ranging with the Recommended BS. When conducting initial ranging to a BS Recommended for Association, the MS uses an allocated unicast ranging opportunity, if available.

The serving BS may buffer incoming data addressed to the MS during the scanning interval and transmit that data after the scanning interval during any interleaving interval or after exit of the Scanning mode. When the Report mode is 0b10 (i.e. event-triggered) in the most recently received MOB_SCN-RSP, the MS scans all the BSs within the Recommended BS list of this message and then transmits a MOB_SCN-REP message to report the scanning results to its serving BS after each scanning period at the time indicated in the MOB_SCN-RSP message. The MS may transmit a MOB_SCN-REP message to report the scanning results to its serving BS at any time. The message will be transmitted on the Primary Management CID.

14.2.3 Association Procedure

Association is an optional initial ranging procedure occurring during the scanning interval with respect to one of the neighbour BSs [2]. The function of Association is to enable the MS to acquire and record ranging parameters and service availability information for the purpose of proper selection of a handover target BS and/or expediting a potential future handover to a target BS. Recorded ranging parameters of an Associated BS may be further used for setting initial ranging values in future ranging events during a handover.

Upon completion of a successful MS initial ranging of a BS, if the RNG-RSP message (sent by the BS) contains a service level prediction parameter set to 2, the MS may mark the BS as Associated in its MS local Association table of identities, recording elements of the RNG-RSP to the MS local Association table and setting an appropriate ageing timer.

There are three levels of Association as follows:

- Association Level 0: Scan/Association without coordination. The serving BS and the MS negotiate the Association duration and intervals (via MOB_SCN-REQ). The serving BS allocates periodic intervals where the MS may range neighbouring BSs. The target BS has no knowledge of the MS. The MS uses the target BS contention-based ranging allocations.
- Association Level 1: Association with coordination. Unilaterally or upon request of the MS (through the MOB_SCN-REQ message), the serving BS provides Association parameters to the MS and coordinates Association between the MS and neighbouring BSs. The target BS reserves a CDMA initial ranging code and an initial ranging slot (transmission opportunity) in a specified dedicated ranging region (rendezvous time). The neighbouring BS may

assign the same code or transmission opportunity to more than one MS, but not both. There is no potential for collision of transmissions from different MSs.

- Association Level 2: network assisted association reporting. The MS may request to perform Association with network assisted Association reporting by sending the MOB_SCN-REQ message, including a list of neighbouring BSs, to the serving BS with scanning type = 0b011. The serving BS may also request this type of Association unilaterally by sending the MOB_SCN-RSP message with the proper indication. The serving BS will then coordinate the Association procedure with the requested neighbouring BSs in a fashion similar to Association Level 1. With Level 2, the MS is only required to transmit the CDMA ranging code to the neighbour BSs. The MS does not wait for RNG-RSP from the neighbour BSs. Instead, the RNG-RSP information on PHY offsets is sent by each neighbour BS to the serving BS over the backbone. The serving BS may aggregate all ranging information into a single MOB_ASC_REPORT, MOB_ASC-REP, Association result REPort, message.

14.2.4 CDMA Handover Ranging and Automatic Adjustment

For OFDMA PHY, 802.16e defines the handover ranging [2]. An MS that wishes to perform handover ranging must take a process similar to that defined in the initial ranging section with the following modifications. In the CDMA handover ranging process, the CDMA handover ranging code is used instead of the initial ranging code. The code is selected from the handover ranging domain. The handover ranging codes are used for ranging with a target BS during the handover. Alternatively, if the BS is pre-notified for the upcoming handover MS, it may provide bandwidth allocation information to the MS using Fast_Ranging_IE to send an RNG-REQ message.

14.3 The Handover Process

The 802.16 standard states that the handover decision algorithm is beyond its scope. The WiMAX Forum documents do not select a handover algorithm either. Only the framework is defined. The MS, using its current information on the neighbour BS or after a request to obtain such information (see the previous section), evaluates its interest in a potential handover with a target BS. Once the handover decision is taken by either the serving BS or the MS, a notification is sent over the MOB_BSHO-REQ (BS Handover REQuest) or the MOB_MSHO-REQ (MS Handover REQuest) MAC management messages, depending on the handover decision maker: the BS or MS. The handover process steps are described in the following.

The handover process is made of five stages which are summarized in Figure 14.1. The HO process stages are described in the following sections [2].

14.3.1 Cell Reselection

Cell reselection [2] refers to the process of an MS scanning and/or association with one or more BS in order to determine their suitability, along with other performance considerations, as a handover target. The MS may use neighbour BS information acquired from a decoded MOB_NBR-ADV message or may make a request to schedule scanning intervals or sleep

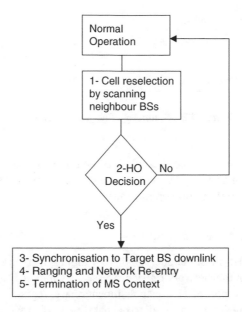

Figure 14.1   Illustration of handover process stages. (Figure by B. Souhaid and L. Nuaymi.)

intervals to scan, and possibly range, the neighbour BS for the purpose of evaluating the MS interest in the handover to a potential target BS.

14.3.2 Handover Decision and Initiation

A handover begins with a decision for an MS to make a handover from a serving BS to a target BS. The decision may originate either at the MS or the serving BS. The handover decision results in a notification of MS intent to make a handover through the MOB_MSHO-REQ (MS HO REQuest) message (handover decision by the MS) or the MOB_BSHO-REQ (BS HO REQuest) message (handover decision by the BS).

The BS may transmit a MOB_BSHO-REQ message when it wants to initiate a handover. This request may be recommended or mandatory. In the case where it is mandatory, at least one recommended BS must be present in the MOB_BSHO-REQ message. If mandatory, the MS responds with the MOB_HO-IND message, indicating commitment to the handover unless the MS is unable to make the handover to any of the recommended BSs in the MOB_ BSHO-REQ message, in which case the MS may respond with the MOB_HO-IND message with proper parameters indicating HO reject. An MS receiving the MOB_BSHO-REQ message may scan recommended neighbour BSs in this message.

In the case of an MS initiated handover, the BS transmits an MOB_BSHO-RSP message upon reception of the MOB_MSHO-REQ message.

14.3.3 Synchronisation to a Target BS Downlink

Synchronisation to a target BS downlink must be done. If the MS had previously received a MOB_NBR-ADV (MAC management) message including a target BSID, physical frequency,

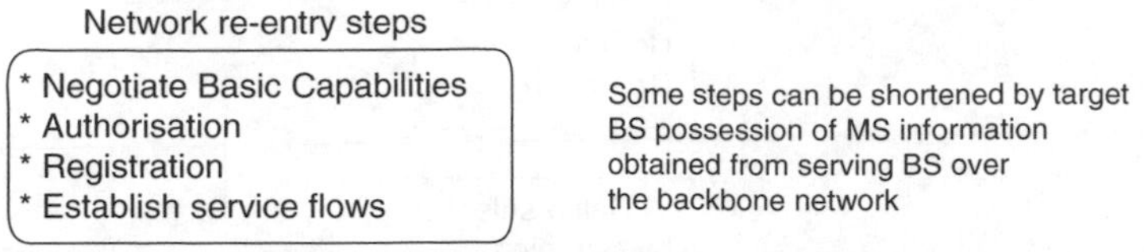

Network re-entry steps

* Negotiate Basic Capabilities   Some steps can be shortened by target
* Authorisation   BS possession of MS information
* Registration   obtained from serving BS over
* Establish service flows   the backbone network

Figure 14.2   Summary of network re-entry steps

DCD and UCD, this process may be shortened. If the target BS had previously received handover notification from a serving BS over the backbone, then the target BS may allocate a non-contention-based initial ranging opportunity.

14.3.4 Ranging and Network Re-entry

The MS and the target BS must conduct handover ranging. Network re-entry proceeds from the initial ranging step in the Network Entry process (see Chapter 11): negotiate basic capabilities, PKM authentication phase, TEK establishment phase, registration (the BS may send an unsolicited REG-RSP message with updated capabilities information or skip the REG-RSP message when there is no TLV information to be updated) and the other following Network Entry optional steps (IP connectivity, etc.).

Network re-entry may be shortened by target BS possession of MS information obtained from the serving BS over the backbone network. Depending on the amount of that information, the target BS may decide to skip one or several of the Network Entry steps (Figure 14.2). Handover ranging can then be a simplified version of initial ranging. To notify an MS seeking handover of possible omission of re-entry process management messages during the current handover attempt (due to the availability of MS service and operational context information obtained over the backbone network), the target BS must place, in the RNG-RSP message, an HO Process Optimisation TLV indicating which re-entry management messages may be omitted. The MS completes the processing of all indicated messages before entering Normal Operation with the target BS.

Regardless of having received MS information from a serving BS, the target BS may request MS information from the backbone network.

14.3.5 Termination of MS Context

This is the final step of a handover. Termination of the MS context is defined as the serving BS termination of the context of all connections belonging to the MS and the discarding of the context associated with them, i.e. information in queues, ARQ state machine, counters, timers, header suppression information, etc. This is accomplished by sending the MOB_HO-IND message with the HO_IND_type value indicating a serving BS release.

14.3.6 Handover Cancellation

An MS may cancel HO at any time prior to expiration of the Resource_Retain_Time interval after transmission of the MOB_HO-IND message. Resource_Retain_Time is one of the parameters exchanged during the registration procedure (part of Network Entry). The standard [2] indicates that Resource_Retain_Time is a multiple of 100 milliseconds and that 200 milliseconds is recommended as default.

14.4 Fast BS Switching (FBSS) and Macro Diversity Handover (MDHO)

14.4.1 Diversity Set

There are several conditions that are required to the diversity BSs featured in FBSS and MDHO procedures. These conditions are listed below [2]:

- The BSs are synchronised based on a common time source and have synchronised frames.
- The frames sent by the BSs from the diversity set arrive at the MS within the prefix interval, i.e. transmission delay < cyclic prefix.
- The BSs operate at same frequency channel.
- The BSs are required to share or transfer MAC context. Such context includes all information MS and BS normally exchange during Network Entry, particularly the authentication state, so that an MS authenticated/registered with one of the BSs from the diversity set BSs is automatically authenticated/registered with other BSs from the same diversity set. The context also includes a set of service flows and corresponding mapping to connections associated with the MS, current authentication and encryption keys associated with the connections. There are also BS conditions specific to MDHO (see below).

An MS may scan the neighbour BSs and then select BSs that are suitable to be included in the diversity set. The MS reports the selected BSs and the diversity set update procedure is performed by the BS and the MS. After an MS or BS has initiated a diversity set update using MOB_MSHO/BSHO-REQ, the MS may cancel the diversity set update at any time. This cancellation is made through transmission of an MOB_HO-IND with proper parameters. The BS may reconfigure the diversity set list and retransmit the MOB_BSHO-RSP message to the MS.

In an MS diversity set, a member identifier, TEMP_BSID, is assigned to each BS in the diversity set.

14.4.2 Different Types of BS for a Given MS

Before getting into the details of make-before-break handover algorithms, FBSS and MDHO, the different types of BS for a given MS are summarized:

- Serving BS. The serving BS is the BS with which the MS has most recently completed registration at the initial Network Entry or during a handover.
- Neighbour BS. A neighbour BS is a BS (other than the serving BS) whose downlink transmission can be (relatively well) received by the MS.
- Target BS. This is the BS that an MS intends to be registered with at the end of a handover.
- Active BS. An active BS is informed of the MS capabilities, security parameters, service flows and full MAC context information. For a Macro Diversity HandOver (MDHO), the MS transmits/receives data to/from all active BSs in the diversity set.
- Anchor BS. For MDHO or FBSS supporting MSs, this is a BS where the MS is registered, synchronised, performs ranging and monitors the downlink for control information (see

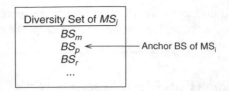

Figure 14.3   Illustration of an anchor BS in a diversity set

Figure 14.3). For an FBSS supporting MS, this is the serving BS that is designated to transmit/receive data to/from the MS at a given frame. Hence, it can be verified that an anchor BS is a specific case of a serving BS. An MS is required continuously to monitor the signal strength of the BSs that are included in the diversity set. The MS selects one BS from its current diversity set to be the anchor BS and reports the selected anchor BS on the CQICH (see Chapter 9) or MOB_MSHO-REQ message. The MSs and BSs may use the fast-feedback method to update the diversity set: when the MS has more than one BS in its diversity set, the MS transmits fast anchor BS selection information to the current anchor BS using the OFDMA fast-feedback channel (see the OFDMA frame in Chapter 9).

14.4.3 FBSS (Fast BS Switching)

An FBSS handover begins with a decision for an MS to receive/transmit data from/to the anchor BS that may change within the diversity set. An FBSS handover can be triggered by either MOB_MSHO-REQ or MOB_BSHO-REQ messages [2].

When operating in FBSS, the MS only communicates with the anchor BS for uplink and downlink messages (management and traffic connections). The MS and BS maintain a list of BSs that are involved in FBSS with the MS. This is the FBSS diversity set. The MS scans the neighbour BSs and selects those that are suitable to be included in the diversity set. Among the BSs in the diversity set, an anchor BS is defined. An FBSS handover is a decision by an MS to receive or transmit data from a new anchor BS within the diversity set.

The MS continuously monitors the signal strength of the BSs of the diversity set and selects one of these BSs to be the anchor BS. Transition from one anchor BS to another, i.e. BS switching, is performed without exchange of explicit handover signalling messages. An important requirement of FBSS is that the data are simultaneously transmitted to all members of a diversity set of BSs that are able to serve the MS.

The FBSS supporting BSs broadcast the DCD message including the H_Add Threshold and H_Delete Threshold. These thresholds may be used by the FBSS-capable MS to determine if MOB_MSHO-REQ should be sent to request switching to another anchor BS or changing diversity set.

14.4.4 MDHO (Macro Diversity Handover)

An MDHO begins with a decision for an MS to transmit to and receive from multiple BSs at the same time. An MDHO can start with either MOB_MSHO-REQ or MOB_BSHO-REQ messages. When operating in an MDHO, the MS communicates with all BSs in the diversity

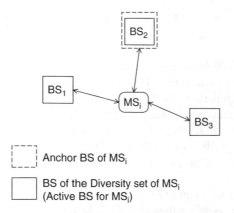

Figure 14.4   Illustration of an MDHO operation mode

set for uplink and downlink unicast traffic messages (see Figure 14.4). The use of this transmission diversity is not the same in the two different communications:

- For a downlink MDHO two or more BSs provide synchronised transmission of MS downlink data such that diversity combining can be performed by the MS.
- For an uplink MDHO, the transmission from an MS is received by multiple BSs such that
  selection diversity of the information received by multiple BSs can be performed.

The BSs involved in an MDHO or equivalently a member of an MS MDHO diversity set
must use the same set of CIDs for the connections that have been established with the MS.
The same MAC/PHY PDUs should be sent by all the BSs involved in the MDHO to the MS.
   The decision to update the diversity set and the process of anchor BS update begin with
notifications by the MS (through the MOB_MSHOREQ message) or by the BS (through the
MOB_BSHO-REQ message).

14.5 Power-Save Modes

IEEE 802.16e defines two new modes: the Sleep mode and the Idle mode in order to have:

- power-efficient MS operation;
- a more efficient handover.

Consequently, the normal operation mode that exists in 802.16-2004 is known as the Active
mode.

14.5.1 Sleep Mode

In the Sleep mode state, the MS conducts pre-negotiated periods of absence from the serving BS air interface. The MS is unavailable to the serving BS (downlink and uplink) in these
periods. The Sleep mode objectives are the following [2]:

- minimise MS power usage;
- minimise the usage of the serving base station air interface resources.

In addition, the MS can scan other base stations to collect information to assist handover during the Sleep mode. Implementation of the Sleep mode is optional for the MS and mandatory for the BS.

For each MS in the Sleep mode, its BS keeps one or several contexts, each one related to a certain Sleep mode power saving class. The power saving class is a group of connections that have common demand properties. There are three types of power saving class, which differ by their parameter sets, procedures of activation/deactivation and policies of MS availability for data transmission. The MOB_SLP-REQ (SLeeP Request Message) (sent by a Sleep mode supporting MS) and the MOB_SLP-RSP (SLeeP Response Message) (sent by the BS) allow a request to be made for a definition and/or activation of certain Sleep mode power-save classes.

The unavailability interval of an MS is a time interval that does not overlap with any listening window of any active power saving class of this MS. During the unavailability interval the BS does not transmit to the MS, so the MS may power down or perform other activities that do not require communication with the BS, such as scanning neighbour BSs, associating with neighbour BSs, etc. During unavailability intervals for the MS, the BS may buffer (or it may drop) MAC SDUs addressed to unicast connections bound to the MS.

14.5.2 Idle Mode

The Idle mode is intended as a mechanism to allow the MS to become periodically available for downlink broadcast traffic messaging without registration at a specific BS as the MS traverses an air link environment populated by multiple BSs, typically over a large geographic area [2]. The Idle mode benefits the MSs by removing the active requirement for handovers and all Active mode normal operation requirements. By restricting MS activity to scanning at discrete intervals, the Idle mode allows the MS to conserve power and operational resources. The Idle mode also benefits the network and the BSs by eliminating air interface and network handover traffic from essentially inactive MSs while still providing a simple and fast method (paging) for alerting the MS about pending downlink traffic.

The BS are divided into logical groups called paging groups. The purpose of these groups is to offer a contiguous coverage region (see Figure 14.5) in which the MS does not need to transmit in the uplink yet can be paged in the downlink if there is traffic targeted at it. The paging groups have to be large enough so that most MSs will remain within the same paging group most of the time and small enough such that the paging overhead is reasonable. A BS may be a member of one or more paging groups.

The MOB_PAG-ADV (BS broadcast PAGing) message is sent by the BS on the Broadcast CID or Idle mode multicast CID during the BS paging interval. This message indicates for a number of Idle mode supporting MSs a requirement to perform ranging to establish location and acknowledge a message or to enter the network. An MS will terminate the Idle mode and re-enter the network if it decodes a MOB_PAG-ADV message that contains the MS MAC address and an action code of 0b10 (Network Entry).

Idle mode initiation may begin after MS de-registration. During the Active mode normal operation with its serving BS, an MS may signal intent to begin the Idle mode by sending a DREG-REQ message with a De-Registration_Request_Code = 0×01, indicating a request for MS de-registration from a serving BS and initiation of the MS Idle mode. At MS Idle mode initiation, an MS may engage in cell selection to obtain a new preferred BS. A preferred

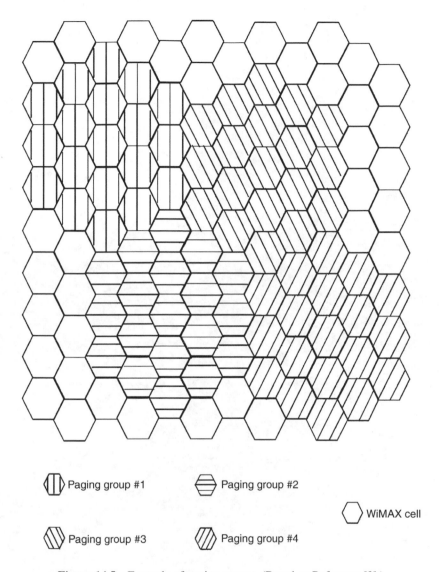

Figure 14.5 Example of paging groups. (Based on Reference [2].)

BS is a neighbour BS that the MS evaluates and selects as the BS with the best air interface downlink properties.

The Idle mode can also be BS initiated: the serving BS includes a REQ duration TLV with an Action code $= 0 \times 05$ in the DREG-CMD message, signalling for an MS to initiate an Idle mode request.

15

Security

15.1 Security Elements Used in the 802.16 Standard

A wireless system uses the radio channel, which is an open channel. Hence, security procedures must be included in order to protect the traffic confidentiality and integrity and to prevent different network security attacks such as theft of service. The IEEE 802.16 MAC layer contains a security sublayer (see the protocol layers in Chapter 3). This sublayer and its main procedures are described in this chapter.

The Privacy Key Management (PKM) protocol, later known as PKMv1, is included in the 802.16-2004 standard security sublayer in order to provide secure distribution of keying data from the BS to the SS. In addition, PKM is used to apply conditional access to network services, making it the authentication protocol, protecting them from theft of service (or service hijacking) and providing a secure key exchange. Many ciphering (data encryption) algorithms are included in the 802.16 standard security sublayer for encrypting packet data across the 802.16 network.

The Security sublayer has been redefined in the IEEE 802.16e amendment mainly due to the fact that 802.16-2004 had some security holes (e.g. no authentication of the BS) and that the security requirements for mobile services are not the same as for fixed services. The Security sublayer has two main component protocols as follows:

- A data encapsulation protocol for securing packet data across the fixed BWA network. This protocol defines a set of supported cryptographic suites, i.e. pairings of data encryption and authentication algorithms, and the rules for applying those algorithms to a MAC PDU payload.
- A key management protocol (PKM) providing the secure distribution of keying data from the BS to the SS. Through this key management protocol, the SSs and BSs synchronise keying data. In addition, the BS uses the protocol to enforce conditional access to network services. The 802.16e amendment defined PKMv2 with enhanced features.

The Security sublayer of WiMAX as it has been redefined in the IEEE 802.16e is shown in Figure 15.1. The elements of this figure will be described in this chapter, where an SS is sometimes denoted MS (as elsewhere in this book).

WiMAX: Technology for Broadband Wireless Access   Loutfi Nuaymi
© 2007 John Wiley & Sons, Ltd

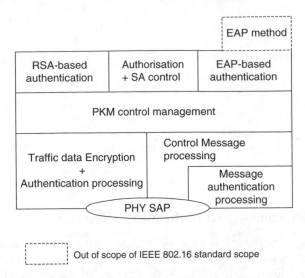

Figure 15.1   Security sublayer components. (Based on Reference [2].)

15.1.1 Encryption Algorithms

Many encryption algorithms are included in the 802.16 standard Security sublayer. They can be used for securing ciphering key exchange and for the encryption of transport data. Some of these algorithms are optional for some applications.

The encryption algorithms included in 802.16 are:

- RSA (Rivest Shamir Adleman) [41]. RSA is a public-key asymmetric encryption algorithm used to encrypt the Authorisation Reply message using the SS public key. The Authorisation Reply message includes the Authorisation Key (AK). RSA may also be used for the encryption of traffic encryption keys when these are transmitted from the BS to the SS.
- DES (Data Encryption Standard) [42]. The DES and 3-DES are shared(secret)-key encryption algorithms. The DES algorithm may be used for traffic data encryption. It is mandatory for 802.16 equipment. The 3-DES algorithm can be used for the encryption of the traffic encryption keys.
- AES (Advanced Encryption Standard) [43]. The AES algorithm is a shared(secret)-key encryption algorithm. The AES algorithm may be used for traffic data encryption and can also be used for the encryption of the traffic encryption keys. Its implementation is optional.

Cryptographic algorithms are also included in 802.16:

- HMAC (Hashed Message Authentication Code) [44, 45] and CMAC (Cipher-based Message Authentication Code) [46]. HMAC and CMAC are used for message authentication and integrity control.

15.1.2 X.509 Certificate

ITU-T X.509 (formerly CCITT X.509) or ISO/IEC 9594-8, which was first published in 1988 as part of the X.500 directory recommendations, defines a standard certificate format [47] used in IETF RFC 3280 [48], itself used in the 802.16 standard (citing RFC 2459 of the IETF).

The 802.16 standard states that 802.16-compliant SSs must use X.509 Version 3 certificate formats providing a public key infrastructure used for secure authentication. Each SS carries a unique X.509 digital certificate issued by the SS manufacturer, known as the SS X.509 certificate. More exactly, this certificate is issued (and signed) by a Certification Authority (CA) and installed by the manufacturer. This digital certificate contains the SS RSA public key and the SS MAC address.

Each SS has a manufacturer-issued X.509 manufacturer CA certificate issued by the manufacturer or by an external authority. The manufacturer's public key is then placed in this X.509 manufacturer CA certificate, which in turn is signed by a higher-level CA. This higher-level CA does not seem to be clearly defined in the present version of the standard. There are then two types of X.509 certificates: SS X.509 certificates and the X.509 manufacturer CA certificate. In the 802.16-2004 standard, there is an X.509 certificate for the BS.

The main fields of the X.509 certificate are the following:

- The X.509 certificate version is always set to v3 when used in the 802.16 standard.
- The certificate serial number is a unique integer the issuing CA assigns to the certificate.
- The signature algorithm is an object identifier and optional parameters defining the algorithm used to sign the certificate.
- The signature value is the digital signature computed on the ASN.1 DER encoded tbsCertificate ('to be signed Certificate'). The standard states that the RSA signature algorithm with SHA-1 (Secure Hash Algorithm), [49] is employed for both defined certificate types.
- The certificate issuer is the Distinguished Name (DN) of the CA that issued the certificate.
- The certificate validity is when the certificate becomes active and when it expires.
- The certificate subject is the DN identifying the entity whose public key is certified in the subject public key information field. The country name, organisation name and manufacturing location are attributes of the certificate subject. Another attribute is the company name for the CA or, for the SS, its 48-bit universal MAC address (see Chapter 8) and its serial number. This MAC address is used to identify the SS to the various provisioning servers during initialisation.
- The Certificate Subject Public Key Info Field contains the public key material (public key and parameters) and the identifier of the algorithm with which the key is used. This is the main content of the X.509 certificate.
- Optional fields allow reuse of issuer names over time.
- The certificate extensions are extension data.

The BS uses the X.509 certificate public key of an SS in order to verify the authenticity of this certificate and then authenticates the SS. This is done using the PKM protocol (see Section 15.2 below). This security information also authenticates the responses received by the SS.

15.1.3 Encryption Keys and Security Associations (SAs)

802.16 standard security uses many encryption keys. The encryption keys defined in 802.16-2004 are listed in Table 15.1 where the notation and the number of bits in each key are given. A nonexhaustive list of the keys used in the PKMv2 protocol, taking into account the 802.16e amendment, is proposed in Table 15.2.

The standard defines a Security Association (SA) as a set of security information a BS and one or more of its client SSs (or MSs) share in order to support secure communications. An

Table 15.1 Encryption keys used in the 802.16 standard, in its 802.16-2004 version, i.e. PKMv1. (Table by L. Rouillé and X. Lagrange at ENST Bretagne.)

Encryption key	Notation	Number of bits	Description
Authorisation Key	AK	160	Authentication of an SS by its BS. Shared secret used to secure further transactions and generating encryption keys.Lifetime between 1 and 70 days
Key Encryption Key	KEK	128	3-DES key used for the encryption of the TEK
Traffic Encryption Key	TEK	128	Data encryption key. Lifetime between 30 min and 7 days
HMAC Key for the Downlink	HMAC_KEY_D	160	Used for authenticating messages in the downlink direction
HMAC Key for the Uplink	HMAC_KEY_U	160	Used for authenticating messages in the uplink direction
HMAC Key in the Mesh mode	HMAC_KEY_S		Used for authenticating messages in the Mesh mode

Table 15.2 The main keys used in the 802.16 standard after the 802.16e amendment. (Table by L. Nuaymi, M. Boutin and M. Jubin.)

Name	Signification	PKM version	Notes
PMK	Pairwise Master Key	2	Obtained from EAP authentication
PAK	Primary Authorisation Key	2	Obtained from RSA-based authorisation
AK	Authorisation Key	1 and 2	Authentication between an SS and a BS. In the case of PKMv2, derived from PMK or PAK
KEK	Key Encryption Key	1 and 2	Used for encryption of TEK
TEK	Traffic Encryption Key	1 and 2	Used for data encryption
GKEK	Group Key Encryption Key	2	Used for encryption of GTEK
GTEK	Group Traffic Encryption Key	2	Used for multicast data packets encryption
MAK	MBS Authorisation Key	2	Authentication for MBS
MGTEK	MBS Group Traffic Encryption Key	2	Used to generate MTK with the MAK
MTK	MBS Traffic Key	2	Protects MBS Traffic. Derived from MAK and MGTEK
H/CMAC_KEY_D		1 and 2	Assures message integrity for the downlink
H/CMAC_KEY_U		1 and 2	Assures message integrity for the uplink

SA's shared information includes the Cryptographic Suite employed within the SA. A Cryptographic Suite is the SA's set of methods for data encryption, data authentication and TEK exchange. The exact content of the SA is dependent on the SA's Cryptographic Suite: encyption keys, keys lifetime, etc. The Security Association Identifier (SAID) is a 16-bit identifier shared between the BS and the SS that uniquely identifies an SA. There are three types of security associations: the SA for unicast connections, the Group Security Association (GSA) for multicast groups and the MBS Group Security Association (MBSGSA) for MBS services.

The SAs are managed by the BS. When an authentication event takes place the BS gives the SS a list of Security Associations associated with its connections. Generally, an SS has a Security Association ('primary' association) for its secondary management connection and two more for the downlink and the uplink links. After that, the BS may indicate one or more new SAs to the SS.

Static SAs are provisioned within the BS. Dynamic SAs are created and deleted as required in response to the initiation and termination of specific service flows.

15.2 Authentication and the PKM Protocol

The security sublayer of IEEE 802.16 employs an authenticated client/server key management protocol in which the BS, the server, controls the distribution of keying material to the client SS. Security is then based on the PKM (Privacy Key Management) protocol [50]. The BS authenticates a client SS during the initial authorisation exchange (see Chapter 11) using digital-certificate-based SS authentication. The PKM protocol uses public key cryptography to establish a shared secret between the SS and the BS. The SS also uses the PKM protocol to support periodic reauthorisation and key refresh.

If during capabilities negotiation (see Section 11.6 for SBC, SS Basic Capability messages exchange) the SS specifies that it does not support IEEE 802.16 security, the authorisation and key exchange procedures in Network Entry process are skipped, i.e. neither key exchange nor data encryption is performed. The standard states that the BS, if provisioned in this way, considers the SS authenticated; otherwise, the SS is not serviced.

The 802.16e amendment defined PKMv2 with enhanced features. Table 15.3 summarises the main differences between the two PKM versions.

First 802.16 PKM protocol MAC management messages are described, followed by the use of these messages and X.509 digital certificates for authentication and key material exchange.

15.2.1 PKM Protocol MAC Management Messages

The PKM protocol has two generic MAC management messages in the 802.16 standard:

- PKM Request (PKM-REQ). The PKM-REQ message encapsulates one PKM message in its message payload. It is always sent from the SS to the BS.
- PKM Response (PKM-RSP). The PKM-RSP message encapsulates one PKM message in its message payload. It is always sent from the BS to the SS.

Both PKM-REQ and PKM-RSP use the primary management connection with the exception that when the BS sends the PKM-RSP message to the SSs for a multicast service or a broadcast service, it may be carried on the broadcast connection. The general formats of PKM-REQ and PKM-RSP messages are shown in Figures 15.2 and 15.3.

Table 15.3   The main differences between PKMv1 and PKMv2. (Table by L. Nuaymi, M. Boutin and M. Jubin.)

Feature	PKMv1	PKMv2
Authentication	RSA-based one-way authentication: the BS authenticates the SS	Mutual authentication. Supports two authentication methods: EAP or RSA
Security Association	One SA family: Unicast.Composed of three types of security associations: primary, dynamic and static	Three SA families: Unicast, Group Security Association and MBS Security Association.Composed of the same three types of security associations as PKMv1
Key encryption	Use of three encryption algorithms: 3-DES, RSA and AES	New encryption method implemented: AES with Key Wrap
Data encryption	Two different algorithms are defined in the standard: • DES in the CBC mode, • AES in the CCM mode	Use of the same algorithms plus AES in the CTR mode and AES in the CBC mode implementation
Other additions	—	Management of security for: • broadcast traffic, • MBS traffic. Definition of a preauthentication procedure in the case of handover

Table 15.4 is a synthesis of 802.16 standard PKMv1 MAC management messages parameters and descriptions. The PKM-REQ and PKM-RSP messages codes added for PKMv2 can be found in Table 26 of Reference [2].

15.2.2 PKMv1: the BS Authenticates the SS and then Provides it with Keying Material

After capability negotiation, the BS authenticates the SS and provides it with key material to enable the ciphering of data. All SSs have factory-installed RSA private/public key pairs or provide an internal algorithm to generate such key pairs dynamically. The SSs with factory-installed RSA key pairs also have factory-installed X.509 certificates. The SSs that rely on internal algorithms to generate an RSA key pair support a mechanism for installing a manufacturer-issued X.509 certificate following key generation.

Thus, each SS has a unique X.509 digital certificate issued by the SS manufacturer [1]. The SS X.509 digital certificate contains the SS public key and the SS MAC address. The

Management Message Type (=9, for PKM-REQ)	Code (8 bits)	PKM Identifier (8 bits)	TLV Encoded Information

Figure 15.2   PKM-REQ MAC management message format

Management Message Type (=10, for PKM-RSP)	Code (8 bits)	PKM Identifier (8 bits)	TLV Encoded Information

Figure 15.3   PKM-RSP MAC management message format

SS X.509 certificate is a public key certificate that binds the SS identifying information to its RSA public key in a verifiable manner. The AK is a secret key obtained from the operator. When requesting an AK, an SS presents its digital certificate to the BS. The SS X.509 certificate is digitally signed by the SS manufacturer and that signature can be verified by a BS that knows the manufacturer's public key.

When requesting an AK (Authorisation Key), an SS presents its X.509 digital certificate and a description of the supported cryptographic algorithms to the BS. The AK is a shared secret used to secure further transactions. The BS verifies the digital certificate, determines the encryption algorithm that should be used and then sends an authentication response to the SS. The verified public key is used to encrypt the AK using the RSA algorithm. The BS then sends this RSA public key encrypted AK to the requesting SS. The procedure of authentication and first key exchange is illustrated in Figure 15.4 and is detailed in the following. The

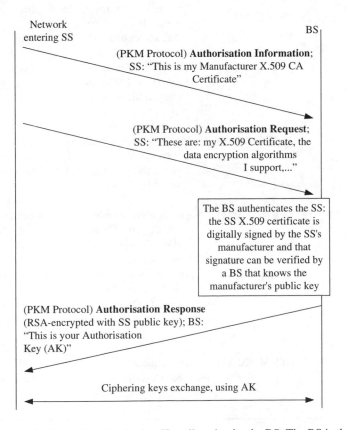

Figure 15.4   Authentication and Authorisation Key allocation by the BS. The BS is the server and the SS is the client

Table 15.4 PKM MAC management messages parameters and descriptions. The Code Field in a PKM-XXX message indicates its type. (Table by L. Rouillé and X. Lagrange at ENST Bretagne.)

PKM protocol message type	MAC message (code)	Message Parameter(s)	Description
Authorisation Info	PKM-REQ (12)	X.509 certificate of a manufacturer CA that issued the SS X.509 certificate	The external CA issues these CA certificates to SS manufacturers
Authorisation Request	PKM-REQ (4)	SS X.509 user certificate, requesting SS security capabilities, SS primary SAID	Request of an AK and a list of SA(s)
Authorisation Reply	PKM-RSP (5)	AK, AK lifetime and sequence number, SA Descriptor(s)	
Authorisation Reject	PKM-RSP (6)	Error Code	The BS rejects the SS Authorisation Request
Authorisation Invalid	PKM-RSP (10)	Error Code identifying reason for Authorisation Invalid (a possibility as no valid AK associated with the requesting SS)	Instructs the receiving SS to reauthorise its BS
Key Request	PKM-REQ (7)	AK sequence number, SAID, HMAC-Digest (Keyed SHA message digest) for message authentication	Request of a TEK for an established SA
Key Reply	PKM-RSP (8)	All of the keying material corresponding to a particular generation of an SAID TEK.	Carries the two TEKs for the concerned SAID. Encrypted (using 3-DES algorithm) with a KEK
Key Reject	PKM-RSP (9)	AK sequence number, SAID, Error Code, HMAC-Digest for message authentication	Indicates the receiving client SS is no longer authorised for a particular SAID
TEK Invalid	PKM-RSP (11)	AK sequence number, SAID, Error Code, HMAC-Digest for message authentication	Indicates to a client SS that the BS determined that the SS encrypted an uplink PDU with an invalid TEK
SA Add	PKM-RSP (3)	AK, sequence number, SA Descriptor(s), HMAC-Digest for message authentication	Establishes one or more additional SAs

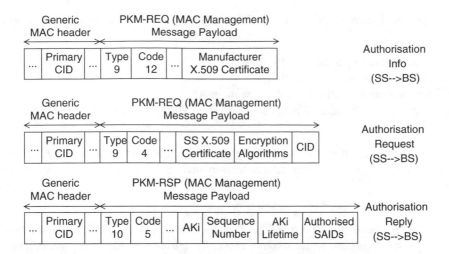

Figure 15.5 MAC management messages as used for PKM protocol-based authentication and key exchange in the 802.16 standard. Some of the parameters are shown. (Figure by L. Rouillé, X. Lagrange and L. Nuaymi.)

above-described PKM MAC management messages are used for authentication and first key exchange (see the illustration in Figure 15.5).

The SS sends a PKM Authorisation Information message. This message contains its manufacturer's X.509 certificate. The Authorisation Information message is strictly an informative message the SS sends to the BS; with it, a BS may dynamically learn the manufacturer's certificate of the client SS.

Then, the SS sends a PKM Authorisation Request message containing the SS X.509 certificate, the SS primary SAID and a description of its security capability (including the ciphering algorithms it supports). The manufacturer's public key is placed in an X.509 Certification Authority (CA) certificate, which in turn is signed by a higher-level CA. The BS verifies the digital certificate and then uses the verified public key to encrypt an AK. The BS then sends this AK to the requesting SS with a PKM Authorisation Response message. The use of the X.509 certificates prevents cloned SSs from passing fake credentials to a BS. The use of a sequence number prevents the so-called replay attack.

As part of their authorisation exchange, the SS provides the BS with a list of all the cryptographic suites (pairing of data encryption and data authentication algorithms) the SS supports. The BS selects from this list a single cryptographic suite to employ with the requesting SS's primary SA and sends this information in the Authorisation Reply message. After this procedure, the SS is required to perform the authentication and key exchange procedures periodically in order to refresh its key material.

It was then pointed out that this one-way authentication (the SS does not authenticate the BS) is a security vulnerability as there can be rogue BSs. This problem is solved with PKMv2.

15.2.3 Mutual Authentication as Defined in 802.16e

The PKMv2 protocol uses the same basis as PKMv1. One of the main differences is the authentication: PKMv2 uses mutual authentication; i.e. the MS also authenticates the BS in order to prevent the connection to a 'false' BS. The BS and MS mutual authentication is the process of:

- the BS authenticating a client's MS identity;
- the MS authenticating the BS identity;
- the BS providing the authenticated MS with an AK;
- the BS providing the authenticated MS with the identities and properties of primary and static SAs.

The key management protocol uses either X.509 digital certificates or EAP for authentication. Therefore, two distinct authentication protocol mechanisms are supported by PKMv2:

- RSA authentication. This protocol is based on X.509 certificates.
- Extensible Authentication Protocol (EAP) authentication (RFC 3748 [39] and RFC 2716 [51]). The EAP uses particular kinds of credential (subscriber identity module, password, token-based, X.509 certificate or other) depending on the EAP method implemented. The EAP methods that are to be used are outside the scope of the standard. Yet, the standard [2] indicates that the EAP method selected should fulfil the 'mandatory criteria' listed in Section 2.2 of RFC 4017 [52]. It seems that, for the moment, EAP-TLS [51] is the only EAP algorithm that fulfils these criterias.

15.2.4 Authorisation Key (AK) Management

The BS is responsible for maintaining the keying information of all its SAs. The standard indicates that the PKMv1 protocol uses public key cryptography to establish a shared secret (i.e. an AK) between the SS and the BS. For the PKMv2 AK generation, see below. The PKM protocol includes a mechanism for synchronising this keying information between a (server) BS and its client SS. This procedure is illustrated in Figure 15.6 and detailed in the following.

As mentioned above, the BS's first receipt of an Authorisation Request message from an unauthorised SS initiates the activation of a new AK, which the BS sends back to the requesting SS in an Authorisation Reply message. This AK remains active until it expires according to its predefined AK lifetime. This parameter is included in the Authorisation Reply message. If an SS fails to reauthorise before the expiration of its current AK, the BS considers this SS unauthorised and then no longer holds an active AK for it.

The SS is responsible for requesting authorisation with its BS and maintaining an active AK. An SS refreshes its AK by reissuing an Authorisation Request to the BS. An AK transition period begins when the BS receives an Authorisation Request message from an SS and the BS has a single active AK for that SS. In response to this Authorisation Request, the BS activates a second AK and sends it back to the requesting SS in an Authorisation Reply message. The BS sets the active lifetime of this second AK to be the remaining lifetime of the first AK, plus the predefined AK lifetime. Thus, the second (newer) key remains active for one AK lifetime beyond the expiration of the first key. Upon achieving authorisation, an SS starts a new encrypted link for each of the SAIDs identified in the Authorisation Reply message. The SS sends a Key Request MAC management message encrypted with the new AK (see Figure 15.6).

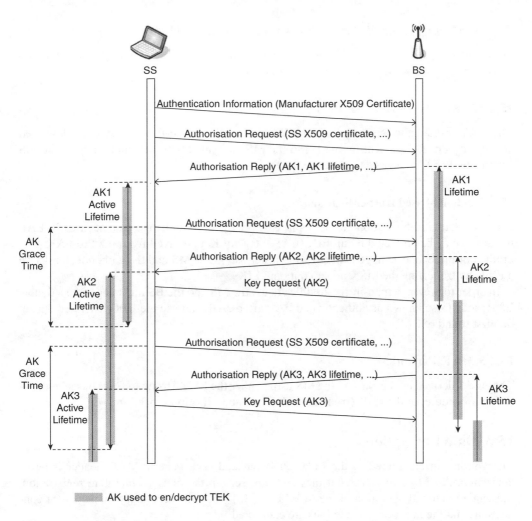

Figure 15.6   Authorisation keys management between the BS and an SS

Two active keys have overlapping lifetimes. The BS is able to support two simultaneously active AKs for each client SS during an AK transition period. The key transition period ends with the expiration of the older key.

An SS schedules the beginning of reauthorisation based on a configurable duration of time, defined in the standard as the Authorisation Grace Time, before the SS's latest AK is scheduled to expire. The Authorisation Grace Time is configured to provide an SS with an authorisation retry period that is sufficiently long to allow for system delays and provide adequate time for the SS to successfully complete an authorisation exchange before the expiration of its most current AK. The BS does not need to know the Authorisation Grace Time. The BS has only to track the lifetimes of its AKs and deactivate each of them when it expires. An AK is used for the generation of encryption keys used for further exchanges, as described in the following section.

15.2.5 Management of the Authorisation Key in PKMv2

Due to PKMv2 there are two different authentication protocols (RSA-based or EAP-based) and two sources of keying materials: RSA and EAP.

15.2.5.1 RSA-based Authorisation

The RSA-based authorisation process yields the Pre-PAK (Pre-Primary AK) as described above. The Pre-PAK is sent by the BS to the MS encrypted with the public key of the MS certificate. Pre-PAK is used to generate the PAK.

15.2.5.2 EAP-based Authentication

An RSA mutual authorisation may take place before the EAP exchange. In that case, the EAP messages may be protected by an EIK (EAP Integrity Key) derived from the Pre-PAK. The product of the EAP exchange is the MSK (Master Key). The MS and the authenticator derive a PMK by truncating the MSK and an optional EIK.

In order to prevent a man-in-the-middle attack, the MS and the BS must negotiate a double EAP mode (also known as authenticated EAP after EAP) and use the EIK generated during the first round of EAP.

15.2.5.3 AK Derivation

The AK is derived by the BS and the MS using either the PMK (from an EAP-based authorisation procedure) or the PAK (from an RSA-based authorisation procedure).

15.3 Data Encryption

Encryption can be applied to the MAC PDU payload. The generic MAC header is never encrypted. All MAC management messages are sent in the clear to facilitate registration, ranging and normal operation of the MAC Layer [2]. Only the secondary management connections and the transport connections are encrypted.

The PKM protocol uses X.509 digital certificates, the RSA public key encryption algorithm and strong encryption algorithms to perform key exchanges between the SS and the BS. After the use of PKM as a public key cryptography to establish a shared secret (the AK) between the SS and the BS, this shared secret is used to secure further PKM exchanges of TEKs (Traffic Encryption Keys). The AK is used for the generation of a Key Encryption Key (KEK) and message authentication keys as now described. The main operation is that the SS

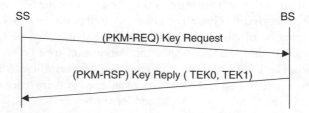

Figure 15.7   The SS asks the BS for encryption keys TEK0 and TEK1

asks the BS for encryption keys (see Figure 15.7). The use of the shared secret, AK, for a two-tiered mechanism for key distribution (described in the following section) permits TEKs to be refreshed without incurring the overhead of computation intensive operations.

15.3.1 Generation of Encryption Keys

At each instant, the BS maintains two sets of active generations of keying material per SA (or, equivalently, per SAID). A set of keying material includes a TEK and its corresponding initialisation vectors (depending of the encryption algorithm). As for the AK, one set corresponds to the older generation of keying material while the second set corresponds to the newer generation of keying material. The BS distributes both generations of active keying material to a client SS. This is done after the SS has requested a TEK with a Key Request message. The Key Reply message contains two TEK parameter attributes, each containing the keying material for one of the SAID's two active sets of keying material.

The BS uses the two active TEKs differently, depending on whether the TEK is used for downlink or uplink traffic. For each of its SAIDs, the BS uses the two active TEKs according to the following rules (illustrated in Figure 15.8):

1. The BS uses the older of the two active TEKs for encrypting downlink traffic.
2. The BS decrypts the uplink traffic using either the older or newer TEK, depending on which of the two keys the SS uses at the time.

Figure 15.9 illustrates the BS management of an SA TEK, where the shaded portion of a TEK's lifetime indicates the time period during which that TEK is used to encrypt MAC PDU payloads. As for AKs, TEKs have a limited lifetime and must be periodically refreshed. At expiration of the older TEK, the BS uses the newer TEK for encryption. It is the responsibility of the SS to update its keys in a timely fashion. The BS uses a KEK when encrypting the TEK in the Key Reply (PKM-RSP) MAC management message. The TEK is encrypted with one of the following algorithms, using the KEK: 3-DES, RSA or AES. The TEK encryption algorithm is indicated by the TEK encryption algorithm identifier in the cryptographic suite of the SA.

A random or pseudo-random number generator is used by the BS to generate KEKs and TEKs. The standard [1] states that recommended practices for generating random numbers for use within cryptographic systems are provided in IETF RFC 1750 [53]. As a remark, it should be pointed out that IETF RFC 1750 has been replaced (by the IETF) by IETF RFC 4086 [54]. For the 3-DES KEK, the 802.16 standard defines the following method [1]:

K_PAD_KEK $= 0 \times 53$ repeated 64 times, i.e. a 512-bit string,
KEK $=$ Truncate(SHA(K_PAD_KEK $|$ AK),128),

where
Truncate (x,n) denotes the result of truncating x to its leftmost n bits,

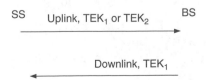

Figure 15.8 Traffic Encryption Keys (TEKs) management. (From IEEE Std 802.16-2004 [1]. Copyright IEEE 2004, IEEE. All rights reserved.)

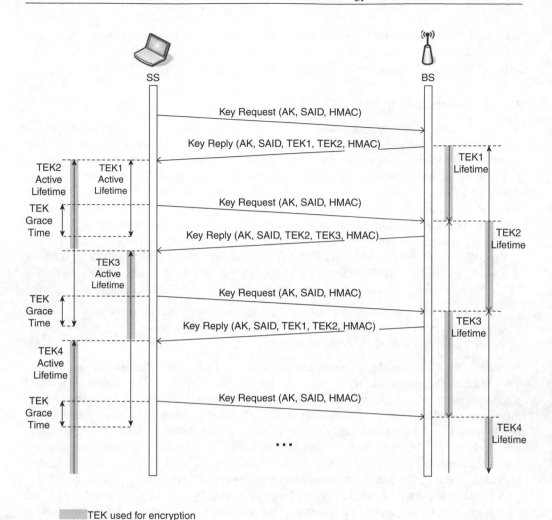

Figure 15.9 Traffic Encryption Keys (TEKs) management

SHA($x|y$) denotes the result of applying the SHA-1 function ([49]) to the concatenated bit strings x and y.

The keying material of 3-DES consists of two distinct DES keys. The 64 most significant bits of the KEK are used in the encrypt operation. The 64 least significant bits are used in the decrypt operation.

The HMAC authentication keys are derived as follows:

HMAC_KEY_D = SHA(H_PAD_D|AK),
HMAC_KEY_U = SHA(H_PAD_U|AK),
HMAC_KEY_S = SHA(H_PAD_D|Operator Shared Secret),

where

H_PAD_D = 0x3A repeated 64 times and H_PAD_U = 0x5C repeated 64 times.

The exchange of AK and the exchange of Traffic Encryption Keys (TEKs) takes place for each SA.

The Operator Shared Secret is a Mesh mode key known by all Mesh mode nodes.

15.3.2 Generation of Encryption Keys in the 802.16e Amendment

As for PKMv1, the PKMv2 protocol uses the TEK and KEK for the encryption of data flows. The KEK is used to encrypt the TEK, GKEK and all other keys sent by the BS to the SS in a unicast message.

15.3.2.1 TEK and KEK

The procedure is the same as for PKMv1 (see the previous section).

15.3.2.2 Multicast Encryption

There is one GKEK per Group Security Association. This GKEK is used to encrypt the GTEKs sent in multicast messages by the BS to the MSs in the same multicast group. It is used to encrypt multicast data packets and it is shared by all the MSs in the multicast group. The GKEK is randomly generated by the BS and encrypted using the same algorithms applied for TEK encryption.

15.3.2.3 Multicast and Broadcast Service Traffic Key

Like the other traffic keys, the MTK is used to encrypt the MBS traffic data. The MGTEK is the GTEK for the MBS. The key generation process in PKMv1 and PKMv2 is summarised in Figure 15.10.

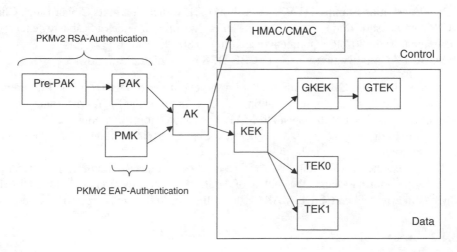

Figure 15.10 Illustration of the key generation process in PKMv1 and PKMv2. (Figure by M. Boutin, M. Jubin and L. Nuaymi.)

15.3.3 Traffic Encryption Keys and Handover

In this section, 802.16e security is considered. The exchange of the SAs and TEK (Traffic Encryption Key) parameters associated with an MS is based on a three-way handshake. This procedure is launched once an authentication event occurs or in the case of a handover:

- During initial network entry or reauthorisation (handover) the BS sends PKMv2 SA-TEK-Challenge.
- The MS sends PKMv2 SA-TEK-Request upon receipt of the PKMv2 SA-TEK-Challenge message.
- To complete the transfer, the BS sends PKMv2-SA-TEK-Response back to the MS.

In the case of a handover, the process can be optimised and the first step may be avoided. Additionally, for each active SA in a previous serving BS, the corresponding TEK, GTEK and GKEK parameters are also included in the new SA. This can be done using the SA-TEK-Update method, which gives the matching between the new SA and the old SA.

15.3.4 Traffic Encryption Algorithms

The MAC PDU payload is encrypted using the active TEK (or GTEK or MTK). The data encryption algorithm is indicated by the data encryption algorithm identifier in the cryptographic suite of an SA. The EC (Encryption Control) bit in the generic MAC header indicates whether the MAC PDU payload is encrypted or not. The generic MAC header is not encrypted. Basic and primary MAC management messages are also not encrypted. However, some of the MAC management messages are authenticated, as described in Section 15.4.

The 802.16-2004 standard included two well-known data encryption algorithms for the encryption of the MAC PDU payloads: DES-CBC and AES-CCM.

15.3.4.1 DES-CBC

The Data Encryption Standard (DES) algorithm ([42]) is included in its Cipher Block Chaining (CBC) mode, shown in Figure 15.11. The DES algorithm provides a 56-bit key encryption and is mandatory in 802.16 equipment (according to 802.16-2004).

The CBC IV (Initialisation Vector) of CBC-DES is calculated as follows:

- In the downlink, the CBC is initialised with the exclusive-or (XOR) of the IV parameter included in the TEK keying information and the content of the PHY synchronisation field (8 bits, right justified) of the latest DL-MAP. The Encryption Key Sequence (EKS), which is a 2-bit field in the generic MAC header (see Chapter 8), is the index of the TEK and IV used to encrypt the payload.
- In the uplink, the CBC is initialised with the XOR of the IV parameter included in the TEK keying information and the content of the PHY synchronisation field of the DL-MAP, which is in effect when the UL-MAP for the uplink transmission is created/received.

15.3.4.2 AES-CCM

The Advanced Encryption Standard (AES) is included in its counter with CBC-MAC (CCM) mode [55]. The AES algorithm is a 128-bit key encryption. The AES is more secure than the

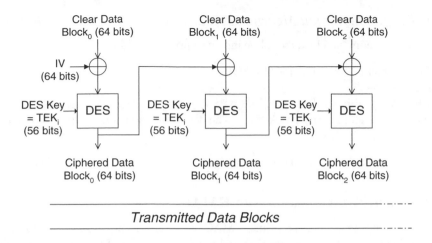

Transmitted Data Blocks

Figure 15.11 DES algorithm in its CBC mode

DES (and the 3-DES also). However, the AES is more complex and a little slower than the DES.

Figure 15.12 shows the AES-CCM generated payload (and then the transported payload). The plaintext PDU is encrypted and authenticated using the active TEK, according to the AES-CCM specification. Before the encrypted data, a 4-byte PN (Packet Number) field is added. This field is linked to the SA and incremented for each PDU transmitted. This is a parade against the replay attack. The replay attack is when some valid packets are replayed by an attacker in order to cause some damage by reproducing an action. Any PN that is received more than once is discarded as a replay attempt. The ciphertext Message Authentication Code, also known as MAC (not to be confused with the Medium Access Layer; see the following section for the Message Authentication Code), is transmitted such that byte index 0 (as enumerated in the AES specification) is transmitted first and byte index 7 is transmitted last (i.e. LSB first). The PN is not encrypted but is included in the message integrity check (Message Authentication Code) calculation.

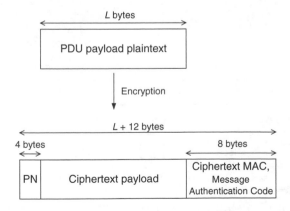

Figure 15.12   Encrypted payload format in the AES-CCM mode. (Based on Reference [2].)

15.3.5 Traffic Encryption Algorithms Added in the 802.16e Amendment

The 802.16e amendment added the following encryption algorithms:

- AES in Counter (CTR) mode [56] for MBS;
- AES in CBC mode;
- AES KeyWrap with a 128-bit key. The AES key wrap encryption algorithm accepts both a ciphertext and an integrity check value. The decryption algorithm returns a plaintext key and the integrity check value.

One or more of the 802.16 defined encryption algorithms can be mandatory in WiMAX profiles.

15.4 Message Authentication with HMAC

The standard states that basic and primary MAC management messages are sent in the clear in order to facilitate registration, ranging and normal operation of the MAC. Thus, authentication and integrity protection of MAC messages is very important. The MAC (Message Authentication Code) sequences, known as keyed hash or MAC-Digest, are used to sign management messages in order to validate their authenticity (see Figure 15.13). The MAC tags are generated and then verified using the same secret key (which is a basic difference with error-detecting codes such as the parity bits). This means that the sender and the receiver of the message must agree on the secret key before starting communications. Evidently, this key is probably different from the encryption key.

The 802.16 standard security includes the use of a Hashed Message Authentication Code (HMAC) for some message authentication and integrity control. 802.16e added the possibility of using CMAC as an alternative to HMAC. The HMAC keyed hash (or HMAC-Digest) is in:

- The HMAC-Digest attribute, which is present in some PKM messages such as Key Request, Key Reply, Key Reject, etc.
- The HMAC Tuple is a TLV parameter used for some MAC management message authentications. The messages that can be authenticated with HMAC include DSx-REQ, DSx-ACK, REG-REQ, etc. The HMAC Tuple is made of HMAC-Digest HMAC with SHA-1, on 160 bits, HMAC Key Sequence Number, on 4 bits, and a reserved field. The HMAC

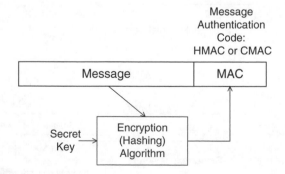

Figure 15.13 Illustration of HMAC or CMAC generation. The MAC is also called keyed hash or MAC Digest

Sequence Number in the HMAC Tuple is equal to the AK Sequence Number of the AK from which the HMAC_KEY_x was derived.

Calculation of the keyed hash in the HMAC-Digest attribute and the HMAC Tuple uses the HMAC [44, 45] with the cryptographic secure hash algorithm, SHA-1 (FIPS 180-1 [49]). This authentication method is often known as HMAC-SHA1. The digest must be calculated over the entire MAC management message with the exception of the HMAC-Digest and HMAC Tuple attributes.

802.16e added the possibility of using a Cipher-based Message Authentication Code (CMAC) (RFC 4493 [46]) as an alternative to the HMAC. For the CMAC, AES block ciphering is used for MAC calculations (AES-CMAC).

The digest is calculated over an entire MAC management message with the exception of the HMAC-Digest or HMAC Tuple attributes.

15.4.1 Message Authentication Keys

The authentication keys used for the calculation of HMAC keyed hash included in some MAC management messages (see above) are:

- the downlink authentication key HMAC_KEY_D used for authenticating messages in the downlink direction;
- the uplink authentication key HMAC_KEY_U used for authenticating messages in the uplink direction.

As for PKMv1, the PKMv2 MAC message for the uplink is C/HMAC_KEY_U and the MAC message for the downlink is C/HMAC_KEY_D. HMAC_KEY_D and HMAC_KEY_U are derived from the AK, as mentioned in Section 15.3 above. The HMAC/CMAC/KEK derivation from the AK is illustrated in Figure 15.14.

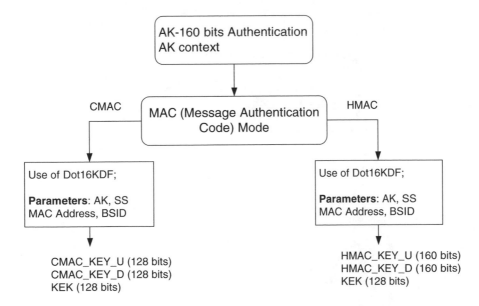

Figure 15.14   HMAC/CMAC/KEK derivation from the AK. (Based on Reference [2].)

The BS uses HMAC_KEY_D and HMAC_KEY_U for the following:

- Verify the HMAC-Digest attributes in Key Request MAC management messages received from that SS, using HMAC_KEY_U.
- Calculate, using HMAC_KEY_D, the HMAC-Digest attributes it writes into Key Reply, Key Reject and TEK Invalid MAC management messages sent to that SS.
- When receiving MAC messages containing the HMAC Tuple attribute, the BS uses the HMAC_KEY_U indicated by the HMAC Key Sequence Number to authenticate the messages.

HMAC_KEY_S is used in the Mesh mode HMAC-Digest calculation.

15.5 Other Security Issues

The procedures seen in this chapter are all about device authentication (SS or BS). Higher-level protocols, such as the higher-level EAP, may be used for this purpose. This type of authentication is part of a WiMAX network specification.

16

Comparisons and Conclusion

16.1 Comparison Between Fixed WiMAX and Mobile WiMAX

In this chapter, some comparisons and then the conclusion are proposed. It is rather risky to give comparisons in a time where the broadband wireless access is at the eve of great changes and innovations. However, based on technical background, many news reports and conference analyses, some comparisons are given. A start is made by comparing Fixed WiMAX and Mobile WiMAX.

Which technology must be chosen? Fixed WiMAX products are already here. The problem is that they can only propose a fixed wireless access, although at rather long distances, up to 20 km. Is it better for an operator to wait some time, until the end of 2007 or the beginning of 2008, according to present expectations, to have Mobile WiMAX? It is up to each operator to decide, taking into account the market targeted. In places where telecommunication infrastructure is well developed, it seems that Fixed WiMAX cannot compete with wired technologies such as DSL. Indeed, it would be surprising to have a wireless (unlimited) Mb/s cheaper than a wired (unlimited) Mb/s in London or Paris one day soon. However, what if this wireless (unlimited) Mb/s includes nomadicity ('your PDA Internet connection works everywhere in the city, although you have to restart your session') and, even more, mobility ('your session is uninterrupted when you move')?

WiMAX has some strong advantages: the same infrastructure can have Fixed and Mobile WiMAX access; the operator can start by covering a small area (if regulatory requirements do not forbid it) in order to adapt the deployment evolution to the business case. This is sometimes known as the 'pay as you grow' model. More generally, the business case must be adapted to the market profile: figures of business travellers, remote (fixed) subscribers, urban technophiles, applications expected (such as Internet, games), etc. This could make, in some cases, Fixed WiMAX a good starter before wide deployments of Mobile WiMAX. This could also give the Fixed WiMAX operator a leading position (reputation, market knowledge, client database, technical teams, etc.) before the deployment of Mobile WiMAX. Mobile WiMAX should normally occupy a majority of the WiMAX landscape for some years. However, a precise estimation of the number of years is thought to be very difficult to give today. This may well leave a market share for Fixed WiMAX, at least for 'some' years. Some applications are, by nature, fixed

WiMAX: Technology for Broadband Wireless Access   Loutfi Nuaymi
© 2007 John Wiley & Sons, Ltd

(e.g. telemetering). On the other hand, it must be kept in mind that Mobile WiMAX can also be used for fixed access from the technical point of view, not taking into account the cost parameter. An important parameter is the spectrum and the cost of this spectrum for each of Fixed and Mobile WiMAX. As of today, these spectrums do not have overlapping zones.

16.2 Comparison Between WiMAX and WiFi

A start can be made by saying that comparing WiMAX and WiFi is comparing two different frameworks. WiMAX has much longer distances and may (or will) also include mobility between cells. In fact, WiFi and WiMAX are complementary, specifically if WiMAX is used for the backhauling of WiFi (see Chapter 1).

There is also a difference in the chronology. WiFi is a WLAN, based on the IEEE 802.11 standard, published in 1997, and the 802.11b variant, published in 1999. WiMAX is a BWA system, including mobility, based on the IEEE 802.16-2004 standard [1], published in 2004, and the 802.16e variant [2], published in February 2006 (in addition to other 802.16 amendments). Hence, if we consider the standard or the products, there is a difference of about six years between the two. In Table 16.1, some comparison elements between WiFi and WiMAX and proposed.

Some precision must be given for the data rate. The one expressed in Table 16.1 is the PHYsical data rate, i.e. the data rate of coded bits. The highest data rate mode is displayed in the table. For all these packet-type transmissions, there is no fixed value for a data layer data rate value due to retransmission, link adaptation, variable header sizes, etc. Standardisation efforts are going on in order to have a higher data rate for IEEE 802.11/WiFi, specifically with the 802.11n variant.

WiMAX has a much better performance than WiFi (range, QoS management, spectrum use efficiency, etc.) but this comes at the price of a higher cost in frequencies and in equipment complexity (and then cost). Consequently, it is definitely not certain that WiMAX could one day soon replace WiFi for some applications.

Table 16.1 Some comparison points between WiFi WLAN and WiMAX BWA

	WiFi/802.11 WLAN	802.16/ WiMAX
Data rate (PHY Layer, optimistic)	54 Mb/s /20 MHz channel	26.2 Mb/s / 7 MHz channel
QoS management	Best Effort, unless for the seldom (until now) implemented 802.11e variant	Five classes of QoS
Multiple access	CSMA/CA (MAC Layer common to 802.11, 802.11a, 802.11b and 802.11g); TDD	TDMA: TDD and FDD. Sophisticated bandwidth reservation mechanisms
Range	Order of magnitude: 100 m	20 km (outdoor CPE), 10 km (indoor CPE)
Frequency bands	Unlicensed	Unlicensed and licensed
Typical use	WLAN	Fixed wireless access, portability, mobility, etc.

16.3 Comparison Between WiMAX and 3G

Table 16.2 gives some comparison elements between major wireless systems: the second-generation cellular system GSM, in its EDGE evolution, 3G UMTS, WiFi in its two variants, 802.11b (the original WiFi) and 802.11a (including OFDM transmission), and WiMAX.

In order to compare with cellular 3G networks, only Mobile WiMAX is considered, since Fixed WiMAX represents a market completely different from 3G. The advantages of each of the two systems are highlighted, starting with the older one, cellular 3G.

16.3.1 Advantages of the 3G Cellular System

- WiMAX uses higher frequencies than Cellular 3G, which mainly operates in the 1.8 GHz range. Received power decreases when frequency increases and wireless system transmitted powers are often limited due to environmental and regulatory requirements. WiMAX ranges are globally smaller than 3G ranges. This is the case for outdoor and indoor equipments. However, the cell range parameter is often not the most limiting one in high-density zones, where the main part of a mobile operator market is located.
- 3G is already here. Its equipment including the high-data rate High-Speed Downlink Packet Access (HSDPA) networks and products are already used, since 2005 in some countries. Globally, 3G has a field advance of two to three years with regard to WiMAX. Will it be enough for 3G to occupy a predominant market share?
- The WiMAX spectrum changes from one country to another. For example, a WiMAX user taking equipment from country A to country B will probably have to use a different WiMAX frequency of the operator of country B. On the other hand, making multifrequency mobile equipment, for a reduced cost, is now becoming more and more easy for manufacturers.

Table 16.2 Some comparison elements between major wireless systems

	Operating frequency	Licensed	One channel (frequency carrier) bandwidth	Number of users per channel	Range
GSM/ EDGE	0.9 GHz, 1.8 GHz, other	Yes	200 kHz	2 to 8	30 km (up to, often less)
UMTS	1.9 GHz	Yes	5 MHz	Many (order of magnitude: 25); data rate decreases	5 km (up to, often less)
WiFi (11b)	2.4 GHz	No	5 MHz	1 (at a given instant)	100 m
WiFi (11a)	5 GHz	No	20 MHz	1 (at a given instant)	100 m
WiMAX	2.3 GHz, 2.5 GHz, 3.5 GHz, 5.8 GHz, other	Licensed and unlicensed bands are defined	3.5 MHz, 7 MHz, 10 MHz, other	Many (100, ...)	20 km (outdoor CPE)

- Some countries have restrictions on WiMAX frequency use, i.e. WiMAX operators can be forbidden to deploy mobility by the regulator.
- Cellular 3G has long had the exclusive support of leading manufacturers, such as Nokia. These companies now seem to be interested in WiMAX while also still remaining very interested in 3G.

16.3.2 Advantages of the (Mobile) WiMAX System

- The frequency spectrum of WiMAX should be cheaper than 3G system frequencies in many countries. The UMTS licence sales in Europe, and specifically in Germany and the UK, reached surprisingly high amounts.
- WiMAX is a very open system as frequently seen in this book: many algorithms are left for the vendor, which opens the door to optimisation, and connections between different business units operating on different parts of the network (core network, radio access network, services providers, etc.), possibly in the same country, are made easy (see Chapter 13). This is probably an advantage, but perhaps it might create some interoperability problems in the first few years?
- The WiMAX PHYsical Layer is based on OFDM, a transmission technique known to have a relatively high spectrum-use efficiency (with regard to SC CDMA). There are plans to upgrade 3G by including OFDM and MIMO in it. This evolution is called, for the moment, LTE (Long-Term Evolution). This gives a time advance for WiMAX in the implementation of OFDM.
- WiMAX is an all-IP technology. This is not the case for the 3G system where many intermediate protocols (tunnelling, etc.) made for the first versions of 3G are not all-IP. However, evolution of 3G should provide end-to-end IP (or all-IP).
- WiMAX has a strong support of some industry giants, such as Intel, KT, Samsung and many others.

Taking into account all these observations, it is very difficult to decide between the two systems. However, if we want to make a guess, it could be said that there is a place for both of these two technologies, depending on the market, the country and the application... at least for a few years to come!

16.4 Final Thoughts and Conclusion

In this book, an attempt has been made to give a global picture of this new and exciting WiMAX technology. WiMAX is based on two sources: the IEEE 802.16 standard, including its amendments, and the WiMAX Forum Group documents. Evidently, this book does not replace these documents, but it is hoped that it will provide a clear introduction to the subject.

WiMAX has a large number of mechanisms and is expected to be used for many applications. The near future will tell which of these mechanisms will be implemented, how they will be implemented and the mechanisms that will be updated by the standardisation bodies.

Annex A

The Different Sets of MAC Management Messages

In this annex, a short description is provided of the main 802.16 MAC management messages. The order of presentation is globally in the order of appearance in the book. In this annex, the broadcast, initial ranging and basic and primary management messages are presented, i.e. all the MAC management messages. It should be remembered that the secondary management messages are upper layer, not MAC, management messages. The MAC management messages cannot be carried on Transport Connections, i.e. with Transport CID values. Full details of these messages can be found in the standards [1] and [2], specifically in Section 6.3.2.3 of the standard and the related PHYsical layers part (Section 8) and TLV encodings (Section 11). The newly added 802.16e messages related to mobility start with MOB.

Multiple access and burst profile definition Messages

Type (8 bits)	Message name	Description	Type of connection
0	UCD, Uplink Channel Descriptor	Transmitted by the BS at a periodic time interval to provide the burst profiles (physical parameters sets) that can be used by an uplink physical channel during a burst in addition to other uplink channel parameters	Broadcast
1	DCD, Downlink Channel Descriptor	Transmitted by the BS at a periodic time interval to provide the burst profiles (physical parameters sets) that can be used by a downlink physical channel during a burst in addition to other downlink channel parameters	Broadcast
2	DL-MAP, Downlink MAP	Downlink access definition. In a DL-MAP, for each downlink burst, DL-MAP_IE indicates the start time and the burst profile (channel details including physical attributes) of this burst	Broadcast
3	UL-MAP, Uplink MAP	Uplink access definition. In a UL-MAP, for each uplink burst, UL-MAP_IE indicates the start time, the duration and the burst profile (channel details including physical attributes) of this burst	Broadcast

WiMAX: Technology for Broadband Wireless Access   Loutfi Nuaymi
© 2007 John Wiley & Sons, Ltd

Mesh network (configuration, entry and scheduling) messages

Type (8 bits)	Message name	Description	Type of connection
39	MSH-NCFG, Mesh Network Configuration	Provides a basic level of communication between nodes in different nearby networks, whether from the same or different equipment vendors or wireless operators. Among others, the Network Descriptor is an embedded data of the MSH-NCFG message. The Network Descriptor contains many channel parameters (modulation and coding schemes, threshold values, etc.) which makes it similar to UCD and DCD	Broadcast
40	MSH-NENT, Mesh Network Entry	Provides the means for a new node to gain synchronisation and initial network entry into a Mesh network	Basic
41	MSH-DSCH, Mesh Distributed Schedule	Transmitted in the Mesh mode when using distributed scheduling. In coordinated distributed scheduling, all the nodes transmit a MSH-DSCH at a regular interval to inform all the neighbours of the schedule of the transmitting station	Broadcast
42	MSH-CSCH, Mesh Centralised Schedule	A MSH-CSCH message is created by a Mesh BS when using centralised scheduling. The BS broadcasts the MSH-CSCH message to all its neighbours and all the nodes with a hop count lower than a given threshold forward the MSH-CSCH message to neighbours that have a higher hop count. This message is used to request or grant bandwidth	Broadcast
43	MSH-CSCF, Mesh Centralised Schedule Configuration	A MSH-CSCF message is broadcasted in the Mesh mode when using centralised scheduling. The Mesh BS broadcasts the MSH-CSCF message to all its neighbours. All nodes forward (rebroadcast) the message according to its index number specified in the message	Broadcast

Management of multicast polling groups messages

Type (8 bits)	Message name	Description	Type of connection
21	MCA-REQ, Multicast Assignment Request	The BS may add (or remove) an SS to a multicast polling group, identified by a multicast CID, by sending an MCA-REQ message with the Join (or leave) command.	Primary management
22	MCA-RSP, Multicast Assignment Response	Sent by the SS in response to an MCA-REQ. Contains mainly the Confirmation Code, indicating whether the request was successful	Primary management

ARQ messages

Type (8 bits)	Message name	Description	Type of connection
33	ARQ-Feedback, Standalone ARQ Feedback message	Standalone ARQ Feedback message. The ARQ-Feedback message can be used to signal any combination of different ARQ ACKs (cumulative, selective, selective with cumulative)	Basic
34	ARQ-Discard, ARQ Discard message	Sent by the transmitter when it wants to skip a certain number of ARQ blocks in the ARQ Transmission Window	Basic
35	ARQ- Reset,ARQ Reset message	Sent by the transmitter or the receiver of an ARQ-enabled transmission in order to reset the parent connection's ARQ transmitter and receiver state machines	Basic

Ranging messages

Type (8 bits)	Message name	Description	Type of connection
4	RNG-REQ, Ranging Request	Transmitted by the SS at initialisation. It can also be used at other periods to determine the network delay and to request a power and/or downlink burst profile change	Initial ranging or Basic
5	RNG-RSP, Ranging Response	Transmitted by the BS in response to a received RNG-REQ. It may also be transmitted asynchronously to send corrections based on measurements that have been made on other received data or MAC messages	Initial ranging or Basic
23	DBPC-REQ, Downlink Burst Profile Change Request	Sent by the SS to the BS on the SS Basic CID to request a change in the downlink burst profile used by the BS to transport data to the SS	Basic
24	DBPC-RSP, Downlink Burst Profile Change Response	Transmitted by the BS on the SS Basic CID in response to a DBPC-REQ message from the SS. If the (required) DIUC parameter is the same as requested in the DBPC-REQ message, then the request was accepted. Otherwise, the DIUC parameter of DBPC-RSP is the previous DIUC at which the SS was receiving downlink data	Basic

Dynamic service management (creation, change and deletion) messages

Type (8 bits)	Message name	Description	Type of connection
11	DSA-REQ, Dynamic Service Addition Request	Sent by an SS or BS to create a new service flow. Service flow attributes, including QoS parameters are indicated	Primary Management
12	DSA-RSP, Dynamic Service Addition Response	Generated in response to a received DSA-REQ; indicates whether the creation of the service flow was successful or rejected	Primary management
13	DSA-ACK, Dynamic Service Addition Acknowledge	Generated in response to a received DSA-RSP	Primary management
14	DSC-REQ, Dynamic Service Change Request	Sent by an SS or BS to change dynamically the parameters of an existing service flow	Primary management
15	DSC-RSP, Dynamic Service Change Response	Generated in response to a received DSC-REQ	Primary management
16	DSC-ACK, Dynamic Service Addition Acknowledge	Generated in response to a received DSC-RSP	Primary management
17	DSD-REQ, Dynamic Service Deletion Request	Sent by an SS or BS to delete an existing service flow	Primary management
18	DSD-RSP, Dynamic Service Deletion Response	Generated in response to a received DSD-REQ	Primary management
30	DSX-RVD, DSx Received Message	Generated by the BS to inform the SS that the BS has correctly received a DSx (DSA or DSC)-REQ message. The DSx-RSP message is transmitted only after the DSx-REQ is authenticated	Primary management

SS basic capability negotiation messages

Type (8 bits)	Message name	Description	Type of connection
26	SBC-REQ, SS Basic Capability Request	Transmitted by the SS during initialisation to inform the BS of its basic capabilities; mainly Physical Parameters and Bandwidth Allocation supported	Basic
27	SBC-RSP, SS Basic Capability Response	Transmitted by the BS in response to an SBC-REQ. Indicates the intersection of the SS and the BS capabilities	Basic

Registration messages

Type (8 bits)	Message name	Description	Type of connection
6	REG-REQ, Registration Request	Sent by the SS in order to register with the BS. Indicates supported management parameters, CS capabilities, IP mode, etc.	Primary management
7	REG-RSP, Registration Response	Sent by the BS in response to a REG-REQ message. Confirms or not authentication. Responds to REG-REQ capability indications	Primary management
29	DREG-CMD, De/ Re-registration Command	Transmitted by the BS to force the SS to change its access state (stop using the current channel, use it again, use it with restrictions, etc.). Unsolicited or in response to an SS DREG-REQ message	Basic
49	DREG-REQ, SS De-registration Request message	Sent by the SS to the BS in order to notify the BS of the SS de-registration request from the BS and the network	Basic

SS reset message

Type (8 bits)	Message name	Description	Type of connection
25	RES-CMD, Reset Command	Transmitted by the BS to force the SS to reset itself, reinitialise its MAC and repeat initial system access	Basic

Configuration file TFTP transmission complete messages

Type (8 bits)	Message name	Description	Type of connection
31	TFTP-CPLT, Config File TFTP Complete message	When the configuration file TFTP download has completed successfully, the SS notifies the BS by transmitting a TFTP-CPLT message	Primary management
32	TFTP-RSP, Config File TFTP Complete Response	In response to TFTP-CPLT, the BS (normally) sends a TFTP-RSP message with an 'OK' response	Primary management

Radio resource management messages

Type (8 bits)	Message name	Description	Type of connection
36	REP-REQ, channel measurement Report Request	Sent by the BS to the SS in order to require RSSI (received power level) and CINR channel measurement reports. In license-exempt bands, the REP-REQ message is also used to request the results of the DFS measurements that the BS has previously scheduled	Basic

(continued overleaf)

(continued)

Type (8 bits)	Message name	Description	Type of connection
37	REP-RSP, channel measurement Report Response	Contains the measurement report in accordance with the Report Request	Basic
38	FPC, Fast Power Control	Broadcast by the BS in order to adjust the power levels of multiple SSs simultaneously. The SSs apply the indicated change within the 'SS downlink management message FPC processing time'. Implementation of the FPC is optional. Power control is normally realised by periodic ranging	Broadcast

Security messages

Type (8 bits)	Message name	Description	Type of connection
9	PKM-REQ, Privacy Key Management Request	Transmits a PKM (Privacy Key Management) protocol message from the PKM client, the SS to the PKM server, the BS	Primary Management
10	PKM-RSP, Privacy Key Management Response	Transmits a PKM protocol message from the PKM server, the BS to the PKM client, the SS	Primary management

AAS (Adaptive Antenna System) messages

Type (8 bits)	Message name	Description (related to AAS operations)	Type of connection
44	AAS-FBCK-REQ	AAS Feedback Request	Basic
45	AAS-FBCK-RSP	AAS Feedback Response	Basic
46	AAS-Beam_Select	AAS Beam Select message	Basic
47	AAS-BEAM_REQ	AAS Beam Request message	Basic
48	AAS-BEAM_RSP	AAS Beam Response message	Basic

Other

Type (8 bits)	Message name	Description	Type of connection
28	CLK-CMP, SS network Clock Comparison	For service flows carrying information that requires the SSs to reconstruct their network clock signals (e.g. Bell-Labs DS1, also known as T1, 1.536 Mb/s circuit transmission system), CLK-CMP messages are periodically broadcast by the BS	Broadcast

A.1 The MAC Management Messages added by 802.16e

The following MAC management messages were defined in 802.16e. They are about mobility, power-save modes, power control, MBS and MIMO.

Power control mode messages

Type (8 bits)	Message name	Description	Type of connection
63	PMC_REQ, Power control Mode Change Request	Sent from the SS to the BS, PMC_REQ is used to request to change the power control mode or to answer PMC_RSP	Basic
64	PMC_RSP, Power control Mode Change Response	The decision of the change of the power control mode (open loop or closed loop) is done at the BS. This decision is indicated by the PMC_RSP MAC message	Basic

MBS message

Type (8 bits)	Message name	Description	Type of connection
62	MBS_MAP	Sent by the BS, on an MBS portion to describe the MBS connections serviced by this MBS portion. If MBS_MAP is not sent, these connections are described in the DL-MAP	—

MIMO precoding setup message

Type (8 bits)	Message name	Description	Type of connection
65	PRC-LT-CTRL, Setup/Tear-down of Long-Term MIMO Precoding	The BS can set up long-term MIMO precoding with feedback from a particular SS by sending this message to this SS. The BS can also use this message to tear down the long-term precoding with feedback	Basic

Handover messages

Type (8 bits)	Message name	Description	Type of connection
53	MOB_NBR-ADV, Neighbour Advertisement	Broadcasted by a BS; provides channel information about neighbouring BSs normally provided by each BS's DCD/UCD message transmissions	Broadcast, Primary management
54	MOB_SCN-REQ, Scanning interval allocation Request	Sent by the SS (MS) to request a scanning interval for the purpose of seeking available BSs and determining their suitability as targets for handover	Basic
55	MOB_SCN-RSP, Scanning interval allocation Response	Sent by the BS to start MS scan reporting with or without scanning allocation	Basic

(continued overleaf)

(continued)

Type (8 bits)	Message name	Description	Type of connection
60	MOB_SCN-REP, Scanning result Report	The MS transmits a MOB_SCN-REP message to report the scanning results to its serving BS after each scanning period at the time indicated in the MOB_SCN-RSP message	Primary management
66	MOB_ASC-REP, Association result Report	When association level 2 (network assisted association reporting) is used, the Serving BS may aggregate all the RNG-RSP messages it receives through the backbone from neighbour BSs into a single MOB_ ASC-REP message, which the Serving BS then sends to the MS	Primary management
56	MOB_BSHO-REQ, BS HO Request	The BS may transmit a MOB_BSHO-REQ message when it wants to initiate a handover	Basic
57	MOB_MSHO-REQ, MS HO Request	The MS may transmit an MOB_MSHO-REQ message when it wants to initiate a handover	Basic
58	MOB_BSHO-RSP, BS HO Response	The BS transmits an MOB_BSHO-RSP message upon reception of an MOB_ MSHO-REQ message	Basic
59	MOB_HO-IND, HO Indication	An MS transmits the MOB_HO-IND message, giving an indication that it is about to perform a handover. When the MS cancels or rejects the handover, it also transmits the MOB_HO-IND message with a proper indication	Basic

Sleep mode messages

Type (8 bits)	Message name	Description	Type of connection
50	MOB_SLP-REQ, Sleep Request	Sent by a Sleep mode supporting MS to request definition and/or activation of certain Sleep mode power-save classes	Basic
51	MOB_SLP-RSP, Sleep Response	Sent from the BS to an MS on Broadcast CID or on the MS Basic CID in response to an MOB_SLP-REQ message, or unsolicited. May contain the definition of a new Sleep Mode Power Saving Class or signal activation	Basic
52	MOB_TRF-IND, Traffic Indication	Sent from the BS to an MS in a Sleep mode that has one or more Sleep Mode Power-Saving Class Type I. This message, sent during those MS listening intervals, indicates whether there has been traffic addressed to any MS that is in Sleep mode	Broadcast

Idle mode message

Type (8 bits)	Message name	Description	Type of connection
61	MOB_PAG-ADV, BS broadcast PAGing	Sent by the BS on the Broadcast CID or Idle mode multicast CID during the BS Paging Interval. Indicates, for a number of Idle mode supporting MSs, the requirement of performing ranging to establish the location, acknowledge a message or to Enter Network	Broadcast

Present versions of the IEEE 802.16g amendment draft include some new MAC management messages. This amendment should be published before the end of the first half of 2007 (October 2006 information).

Annex B

Example of the Downlink Channel Descriptor (DCD) Message

In Chapter 9, were given the global parts of a DCD message, a MAC management message that describes the PHY characteristics of the downlink channel. This example of a DCD message contains the descriptions of two downlink burst profiles respectively identified by DIUC 0101 (hexadecimal: 5) and DIUC 1010 (hexadecimal: A). In this Annex the full details of this message are given, using OFDM (WiMAX) PHYsical interface specifications.

The construction of this message is based on Tables 233 and 358 and, more generally, on Section 11.4 of the IEEE 802.16-2004 specification. Some of the numerical values chosen may not be completely based on practical considerations (e.g. BSID, whose definition is not exhaustive in the standard). 802.16e added new parameters to the DCD message (mainly for the handover process) that are not included in this example.

Field content	Field length	Field name	Description
01	8b (fixed)	MAC Management Message Type	01 is the value for DCD
04	8b (fixed)	Downlink Channel ID	Arbitrarily chosen by the BS
01	8b (fixed)	Configuration Change Count	Indicates a change versus previous DCD message (01 is an assumption)

Start of the Downlink Burst Profile Description for **DIUC value = 0101** (hexadecimal:5). This part is TLV encoded.

01	8b (fixed)	Downlink_Burst_Profile Type	Downlink_Burst_Profile Indicator
44 (decimal: 67)	8b (fixed)	Downlink_Burst_Profile Length	67 bytes for the overal Downlink_Burst_ Profile of DIUC = 0101
0	4b (fixed)	*Reserved*	Set to zero
05	4b (fixed)	*DIUC* value	**This value of DIUC will be associated with the Downlink Burst Profile and Thresholds defined in the following**

Start of Channel Encodings values for DIUC value = 0101 (hexadecimal:5). This part is TLV encoded.

01	1 byte (fixed)	Type for **Downlink_Burst_ Profile** encoding	The number of bytes in the overall object, including embedded TLV items
01	1 byte	Length for Downlink_Burst_ Profile encoding	See Rule for TLV Length
44 (decimal: 67)	1 byte	Value for Downlink_Burst_ Profile encoding	67 bytes for the overall Downlink_ Burst_Profile of DIUC =0101
02	1 byte (fixed)	Type for **BS EIRP** encoding	The BS Transmitted Power (effective isotropic radiated power)
01	1 byte	Length for BS EIRP encoding	See Rule for TLV Length
1E	1 byte	Value for BS EIRP encoding	Signed in units of 1 dBm (see Note 1 at the end of annex B)
06	1 byte (fixed)	Type for **Downlink channel number** encoding	Used for license-exempt operation only
01	1 byte	Length for Downlink channel number encoding	
09	1 byte (fixed)	Value for Downlink channel number encoding	The Channel Number (Channel Nr), an 8-bit value, allows the calculation of the channel centre frequency for license-exempt frequency bands
07	1 byte (fixed)	Type for **TTG** (Transmit/ receive Transition Gap) encoding	Expressed in Physical Slot
01	1 byte	Length for TTG (Transmit/ receive Transition Gap) encoding	
2D	1 byte	Value for TTG (Transmit/ receive Transition Gap) encoding	148 Physical Slots
08	1 byte (fixed)	Type for **RTG** (Receive/transmit Transition Gap) encoding	Expressed in Physical Slot
01	1 byte	Length for RTG (Receive/transmit Transition Gap) encoding	
2D	1 byte	Value for RTG (Receive/transmit Transition Gap) encoding	84 Physical Slots
09	1 byte (fixed)	Type for **Initial Ranging Maximal Received Signal Strength at BS** encoding	$RSS_{IR,max}$: Initial Ranging Maximal Received Signal Strength at BS
02	1 byte	Length for Initial Ranging Maximal Received Signal Strength at BS encoding	
FFE4	2 bytes	Value for Initial Ranging Maximal Received Signal Strength at BS encoding	Signed in units of 1 dBm (see Note 1 below). FFE4 = -28 dBm ($10^{-5.8}$ W)
A (decimal: 10)	1 byte (fixed)	Type for **Channel Switch Frame Number** encoding	Used for license-exempt operation only. In the case of DFS (Dynamic Frequency Selection), the new channel to be used is deduced by using this Channel Switch Frame Number

(continued)

03	1 byte	Length for Channel Switch Frame Number encoding	See Rule for TLV Length
000003	3 bytes	Value for Channel Switch Frame Number encoding	+3
C (decimal: 12)	1 byte (fixed)	Type for **Downlink centre frequency** encoding	The centre of the frequency band in which a base station (BS) or SS is intended to transmit in kHz
04	1 byte	Length for Downlink centre frequency encoding	
0036 5240	4 bytes	Value for Downlink centre frequency encoding	00365240 (0000,0000,0011,0110, 0101,1111,1110,1010) =3 560 000 kHz (3.56 GHz)
D (decimal: 13)	1 byte (fixed)	Type for **Base Station ID** encoding	The Base Station ID is a 48-bit long field identifying the BS (not the MAC address)
06	1 byte	Length for Base Station ID encoding	
AE54AA 123456	6 bytes	Value for Base Station IDencoding	The Base Station ID is programmable. The MS 24 bits must be used as the **operator ID**. A random value has been chosen here
E (decimal: 14)	1 byte (fixed)	Type for **Frame Duration Code** encoding	The Frame Duration Code values are given in the standard
01	1 byte	Length for Frame Duration Code encoding	
03	1 bytes	Value for Frame Duration Code encoding	The code corresponding to 8 ms (see Table 232 of 802.16-2004)
F (decimal: 15)	1 byte (fixed)	Type for **Frame Number** encoding	The Frame Number is a modulo 2^{12} number (a 12-bit-number) which is increased by one for every frame
03	1 byte	Length for Frame Number encoding	
A0B1EE	3 bytes	Value for Frame Number encoding	The frame number of the frame containing the DCD message
94h (decimal: 148)	1 byte (fixed)	Type for **MAC version** encoding	Specifies the version of IEEE 802.16 to which the message originator conforms
01	1 byte	Length for MAC version encoding	
04	1 byte	Value for MAC version encoding	Indicates conformance with IEEE Std 802.16-2004

Start of Burst Profile Encoding values for DIUC value = 0101 (hexadecimal:5). This part is TLV encoded.

01	1 byte (fixed)	Type for **Frequency** encoding	Downlink frequency of this burst profile in kHz
04	1 byte	Length for Frequency encoding	
0036 5FEA	4 bytes	Value for Frequency encoding	00365FEA (0000,0000,0011, 0110, 0101,1111,1110,1010) = 3 563 500 kHz (3.5635 GHz)
96 (decimal: 150)	1 byte (fixed)	Type for **FEC Code** encoding	Provides the modulation and the coding scheme of the burst profile
01	1 byte	Length for FEC Code encoding	
03	1 byte	Value for FEC Code encoding	This value corresponds to **16-QAM** modulation and Reed–Solomon CC, **coding rate = 1/2**
97 (decimal: 151)	1 byte (fixed)	Type for **DIUC mandatory exit threshold** encoding	CINR (Carrier-to-Interference-and-Noise Ratio) used for burst profile (or equivalently DIUC) selection. In 0.25 dB units.
01	1 byte	Length for DIUC mandatory exit threshold encoding	
30 (decimal: 48)	1 byte	Value for DIUC mandatory exit threshold encoding	Corresponds to 48×0.25 dB = 12 dB
98 (decimal: 152)	1 byte (fixed)	Type for **DIUC mandatory entry threshold** encoding	CINR (Carrier-to-Interference-and-Noise Ratio) used for burst profile (or equivalently DIUC) selection. In 0.25 dB units.
01	1 byte	Length for DIUC mandatory entry threshold encoding	
36 (decimal: 54)	1 byte	Value for DIUC mandatory entry threshold encoding	Corresponds to 54×0.25 dB = 13.5 dB
99 (decimal: 153)	1 byte (fixed)	Type for **TCS_enable** encoding	The TCS (Transmission Convergence Sublayer) is an optional mechanism for the OFDM PHY(8.1.4.3) 0 = TCS disabled 1 = TCS enabled 2–255 = *Reserved*
01	1 byte	Length for TCS_enable encoding	
1	1 byte	Value for TCS_enable encoding	1 = TCS enabled

Start of Downlink Burst Profile Description **for DIUC value = 1010** (hexadecimal:A). This part is TLV encoded.

01	8b (fixed)	Downlink_Burst_Profile Type	Downlink_Burst_Profile Indicator
44 (decimal: 67)	8b (fixed)	Downlink_Burst_Profile Length	67 bytes for the overal Downlink_Burst_ Profile of DIUC = 1010
0	4b (fixed)	*Reserved*	Set to zero
A	4b (fixed)	DIUC value	**This value of DIUC will be associated with the Downlink Burst Profile and Thresholds defined in the following**

Start of Channel Encodings values for DIUC value = 1010 (hexadecimal:A). This part is TLV encoded.

01	1 byte (fixed)	Type for **Downlink_Burst_ Profile** encoding	The number of bytes in the overall object, including embedded TLV items
01	1 byte	Length for Downlink_Burst_ Profile encoding	See Rule for TLV Length
44 (decimal: 67)	1 byte	Value for Downlink_Burst_ Profile encoding	67 bytes for the overall Downlink_ Burst_Profile of DIUC = 1010
02	1 byte (fixed)	Type for **BS EIRP** encoding	The BS Transmitted Power (effective isotropic radiated power)
01	1 byte	Length for BS EIRP encoding	See Rule for TLV Length
1E	1 byte	Value for BS EIRP encoding	Signed in units of 1 dBm (see Note 1 below). 1E = 30 dBm (1 W)
06	1 byte (fixed)	Type for **Downlink channel number** encoding	Used for license-exempt operation only
01	1 byte	Length for Downlink channel number encoding	
09	1 byte (fixed)	Value for Downlink channel number encoding	The Channel Number (Channel Nr), an 8-bit value, allows the calculation of the channel centre frequency for license-exempt frequency bands
07	1 byte (fixed)	Type for **TTG** (Transmit/ receive Transition Gap) encoding	Expressed in Physical Slot
01	1 byte	Length for TTG (Transmit/ receive Transition Gap) encoding	
3C	1 byte	Value for TTG (Transmit/ receive Transition Gap) encoding	148 Physical Slots
08	1 byte (fixed)	Type for **RTG** (Receive/ transmit Transition Gap) encoding	Expressed in Physical Slot
01	1 byte	Length for RTG (Receive/ transmit Transition Gap) encoding	
3C	1 byte	Value for RTG (Receive/ transmit Transition Gap) encoding	84 Physical Slots
09	1 byte (fixed)	Type for **Initial Ranging Maximal Received Signal Strength at BS** encoding	$RSS_{IR,max}$: Initial Ranging Maximal Received Signal Strength at the BS
02	1 byte	Length for Initial Ranging Maximal Received Signal Strength at BS encoding	

(continued overleaf)

(*continued*)

FFE6	2 bytes	Value for Initial Ranging Maximal Received Signal Strength at BS encoding	Signed in units of 1 dBm (see Note 1 below). FFE6 = −26 dBm ($10^{-5.6}$ W)
A (decimal: 10)	1 byte (fixed)	Type for **Channel Switch Frame Number** encoding	Used for license-exempt operation only. In the case of DFS (Dynamic Frequency Selection), the new channel to be used is deduced by using this Channel Switch Frame Number
03	1 byte	Length for Channel Switch Frame Number encoding	See Rule for TLV Length
000003	3 bytes	Value for Channel Switch Frame Number encoding	+3
C (decimal: 12)	1 byte (fixed)	Type for **Downlink centre frequency** encoding	The centre of the frequency band in which a base station (BS) or SS is intended to transmit in kHz
04	1 byte	Length for Downlink centre frequency encoding	
0036 5240	4 bytes	Value for Downlink centre frequency encoding	00365240 (0000,0000,0011,0110, 0101,1111,1110,1010) = 3 560 000 kHz (3.56 GHz)
D (decimal: 13)	1 byte (fixed)	Type for **Base Station ID** encoding	The Base Station ID is a 48-bit long field identifying the BS (not the MAC address)
06	1 byte	Length for Base Station ID encoding	
AE54 123456	6 bytes	Value for Base Station ID encoding	The Base Station ID is programmable. The MS 24 bits must be used as the operator ID. A random value has been chosen here
E (decimal: 14)	1 byte (fixed)	Type for **Frame Duration Code** encoding	The Frame Duration Code values are given in the standard
01	1 byte	Length for Frame Duration Code encoding	
04	1 byte	Value for Frame Duration Code encoding	The code corresponding to 10 ms (see Table 232 of 802.16-2004)
F (decimal: 15)	1 byte (fixed)	Type for **Frame Number** encoding	The Frame Number is a modulo 2^{12} number (a 12-bit-number) which is increased by one for every frame
03	1 byte	Length for Frame Number encoding	
A0B1EE	3 bytes	Value for Frame Number encoding	The Frame Number of the frame containing the DCD message
94h (decimal: 148)	1 byte (fixed)	Type for **MAC version** encoding	Specifies the version of IEEE 802.16 to which the message originator conforms
01	1 byte	Length for MAC version encoding	
04	1 byte	Value for MAC version encoding	Indicates conformance with IEEE Std 802.16-2004

Start of Burst Profile Encoding values for DIUC value = 1010 (hexadecimal:A). This part is TLV encoded.

01	1 byte (fixed)	Type for **Frequency** encoding	Downlink frequency of this burst profile in kHz
04	1 byte	Length for Frequency encoding	
0036 5FEA	4 bytes	Value for Frequency encoding	00365FEA (0000,0000,0011, 0110,0101,1111,1110,1010) = 3 563 500 kHz (3.5635 GHz)
96 (decimal: 150)	1 byte (fixed)	Type for **FEC Code** encoding	Provides the modulation and the coding scheme of the burst profile
01	1 byte	Length for FEC Code encoding	
05	1 byte	Value for FEC Code encoding	This value corresponds to 64-QAM modulation and Reed–Solomon CC, coding rate = 2/3
97 (decimal: 151)	1 byte (fixed)	Type for **DIUC mandatory exit threshold** encoding	CINR (Carrier-to-Interference-and-Noise Ratio) used for burst profile (or equivalently DIUC) selection. In 0.25 dB units
01	1 byte	Length for DIUC mandatory exit threshold encoding	
52 (decimal: 82)	1 byte	Value for DIUC mandatory exit threshold encoding	Corresponds to 82 × 0.25 dB = 20.5 dB
98 (decimal: 152)	1 byte (fixed)	Type for **DIUC mandatory entry threshold** encoding	CINR (Carrier-to-Interference-and-Noise Ratio) used for burst profile (or equivalently DIUC) selection. In 0.25 dB units
01	1 byte	Length for DIUC mandatory entry threshold encoding	
62 (decimal: 98)	1 byte	Value for DIUC mandatory entry threshold encoding	Corresponds to 98 × 0.25 dB = 24.5 dB
99 (decimal: 153)	1 byte (fixed)	Type for **TCS_enable** encoding	The TCS (Transmission Convergence Sublayer) is an optional mechanism for the OFDM PHY 0 = TCS disabled 1 = TCS enabled 2–255 = *Reserved*
01	1 byte	Length for TCS_enable encoding	
1	1 byte	Value for TCS_enable encoding	1 = TCS enabled

Note 1. No specific rule is given in the standard (802.16-2004) for some parameters coding. It is presumed that for N possible combinations (e.g. for 16 bits, 65 536 combinations), the first $N/2$ combinations (0 to 32 767) represent positive numbers while the next 32 768 combinations (32 768 to 65 536) represent negative numbers (two's complement). In order to get '−28', 28 is subtracted from the number of combinations (65 536) in order to get 65 508 (binary: 1111,1111,1110,0100, hexadecimal: FFE4).

References

[1] IEEE 802.16-2004, *IEEE Standard for Local and Metropolitan Area Networks, Air Interface for Fixed Broadband Wireless Access Systems*, October 2004.

[2] IEEE 802.16e, *IEEE Standard for Local and Metropolitan Area Networks, Air Interface for Fixed Broadband Wireless Access Systems*, Amendment 2: *Physical and Medium Access Control Layers for Combined Fixed and Mobile Operation in Licensed Bands and Corrigendum 1*, February 2006 (Approved: 7 December 2005).

[3] *Wikipedia*, the free encyclopedia, www.wikipedia.org.

[4] Tanenbaum, A. *Computer Networks*, Prentice-Hall, August 2002.

[5] Agis, A. *et al.*, Global, interoperable broadband wireless networks: extending WiMAX technology to mobility. *Intel Technology Journal*, August 2004.

[6] WiMAX Forum White Paper, 3rd WiMAX Forum plugfest – test methodology and key learnings, March 2006.

[7] Recommendation ITU-R M.1652, Dynamic frequency selection (DFS) in wireless access systems including radio local area networks for the purpose of protecting the radio determination service in the 5 GHz band, 2003.

[8] IEEE 802.16f, *IEEE Standard for Local and Metropolitan Area Networks, Air Interface for Fixed Broadband Wireless Access Systems*, Amendment 1: *Management Information Base*, December 2005.

[9] Lee, W. C. Y., *Mobile Cellular Telecommunications: Analog and Digital Systems*, McGraw-Hill, 2000.

[10] WiMAX Forum White Paper, Mobile WiMAX – Part I: a technical overview and performance evaluation, March 2006.

[11] WiMAX Forum White Paper, Initial certification profiles and the European regulatory framework, September 2004.

[12] van Nee, R. and Prasad, R., *OFDM for Wireless Multimedia Communications*, Artech House, 2000.

[13] Holma, H. and Toksala, A., *WCDMA for UMTS*, 3rd edn, John Wiley & Sons, Ltd, 2004.

[14] IETF RFC 3095, *RObust Header Compression (ROHC): Framework and Four Profiles: RTP, UDP, ESP, and Uncompressed*, C. Bormann *et al.*, July 2001.

[15] IETF RFC 3545, *Enhanced Compressed RTP (CRTP) for Links with High Delay, Packet Loss and Reordering*, T. Koren *et al.*, July 2003.

[16] Johnston, D. and Yaghoobi, H., Peering into the WiMAX spec, CommsDesign (http://www.commsdesign.com/), January 2004.

[17] IETF RFC 2131, *Dynamic Host Configuration Protocol (DHCP)*, R. Droms, March 1997.

[18] IETF RFC 2132, *DHCP options and BOOTP Vendor Extensions*, S. Alexander and R. Droms, March 1997.

[19] IETF RFC 868, *Time Protocol*, J. Postel and K. Harrenstien, May 1983.

[20] IETF RFC 1350, *The TFTP Protocol (Revision 2)*, K. Sollins, July 1992.

[21] WiMAX Forum Document, WiMAX end-to-end network systems architecture; Stage 2, Release 1: architecture tenets, network reference architecture, reference points, April 2006.

[22] Rappaport, T.S., *Wireless Communications; Principles and Practice*, Prentice-Hall, 1996.

[23] Recommendation ITU-R P.526-8, Propagation by diffraction, 2003.

[24] Erceg, V., *et al.*, Channel models for fixed wireless applications, IEEE 802.16 Broadband Wireless Access working group, February 2001.

[25] Recommendation ITU-R P.530-9, Propagation data and prediction methods required for the design of terrestrial line-of-sight systems, 2001.

[26] WiMAX Forum White Paper, WiMAX deployment considerations for fixed wireless access in the 2.5 GHz and 3.5 GHz licensed bands, June 05.

[27] Lehne, P.H. and Pettersen, M., An overview of smart antenna technology for mobile communications systems. *IEEE Communications Surveys and Tutorials*, **2**(4), Fourth Quarter 1999.

[28] Liberti, J.C. and Rappaport, T.S., *Smart Antennas for Wireless Communications: IS-95 and Third Generation WCDMA Applications*, Prentice-Hall, 1999.

[29] Kuchar, A., Taferner, M., Tangemann, M. and Hoek, C., Field trial with GSM/DCS1800 smart antenna base station. Proceedings of the Vehicular Technology Conference, VTC Fall, September 1999.

[30] Alamouti, S., A simple transmit diversity technique for wireless communications. *IEEE Journal on Selected Areas in Communications*, **16**(8), October 1998.

[31] Paulraj, A.J. and Papadias, C.B., Space–time processing for wireless communications. *IEEE Signal Processing Magazine*, November 1997.

[32] Gesbert, D., Shafi, M., Shiu, D.S., Smith, P.J. and Naguib, A., From theory to practice: an overview of MIMO space–time coded wireless systems. *IEEE Journal on Selected Areas in Communications*, **21**(3), April 1998.

[33] WiMAX Forum White Paper, Fixed, nomadic, portable and mobile applications for 802.16-2004 and 802.16e WiMAX networks, November 2005.

[34] WiMAX Forum Document, WiMAX end-to-end network systems architecture; Stage 3, Release 1: detailed protocols and procedures, April 2006.

[35] IETF RFC 1701, *Generic Encapsulation Protocol (GRE)*, S. Hanks *et al.*, October 1994.

[36] IETF RFC 3315, *Dynamic Host Configuration Protocol for IPv6 (DHCPv6)*, R. Droms *et al.*, July 2003.

[37] IETF RFC 2462, *IPv6 Stateless Address Autoconfiguration, IETF Standard*, S. Thomson *et al.*, December 1998.

[38] IETF RFC 3041, *Privacy Extensions for Stateless Address Auto-configuration in IPv6*, T. Narten and R. Draves, January 2001.

[39] IETF RFC 3748, *Extensible Authentication Protocol (EAP)*, B. Aboba *et al.*, June 2004.

[40] IETF RFC 3579, *RADIUS (Remote Authentication Dial In User Service) Support for Extensible Authentication Protocol (EAP)*, B. Aboba and P. Calhoun, September 2003.

[41] PKCS #1 v2.0, *RSA Cryptography Standard*, RSA Laboratories, October 1998.

[42] National Technical Information Service (NTIS), FIPS 46-3, *Data Encryption Standard (DES)*, October 1999; http://www.ntis.gov/.

[43] National Technical Information Service (NTIS), FIPS 197, *Advanced Encryption Standard (AES)*, November 2001.

[44] National Technical Information Service (NTIS), FIPS 198, *The Keyed-Hash Message Authentication Code (HMAC)*, March 2002.

[45] IETF RFC 2104, *HMAC: Keyed-Hashing for Message Authentication*, H. Krawczyk, M. Bellare and R. Canetti, February 1997.

[46] IETF RFC 4493, *The AES-CMAC Algorithm*, J.H. Song *et al.*, June 2006.

[47] ITU-T Recommendation X.509 (1997 E), Information technology – open systems interconnection – the directory: authentication framework, June 1997.

[48] IETF RFC 3280, *Internet X.509 Public Key Infrastructure Certificate and Certificate Revocation List (CRL) Profile* (eds R. Housley *et al.*,) April 2002.

[49] National Technical Information Service (NTIS), FIPS 180-1, *Secure Hash Standard (SHA-1)*, April 1995.

[50] SCTE DSS 00-09, Data over cable service interface specification baseline privacy interface, DOCSIS SP-BPI+-I06-001215, Baseline privacy plus interface specification, December 2000.

[51] IETF RFC 2716, *PPP EAP TLS Authentication Protocol*, B. Aboba and D. Simon, October 1999.

[52] IETF RFC 4017, *Extensible Authentication Protocol (EAP) Method Requirements for Wireless LANs*, D. Stanley *et al.*, March 2005.

[53] IETF RFC 1750, *Randomness Recommendations for Security*, D. Eastlake *et al.*, December 1994.

[54] IETF RFC 4086, *Randomness Requirements for Security*, D. Eastlake *et al.*, June 2005.

[55] IETF RFC 3610, *Counter with CBC-MAC (CCM)*, D. Whiting *et al.*, September 2003.

[56] IETF RFC 3686, *Using Advanced Encryption Standard (AES) Counter Mode with IPsec Encapsulating Security Payload (ESP)*, R. Housley *et al.*, January 2004.

Index

16-QAM, *see* QAM
3G cellular systems, 1, 253–4
64-QAM, *see* QAM

802.16e, 14
802.16f, 20, 28
802.16g, 20
802.16j, 21

AAA, 208, 216–7
AAS, 14, 195–9
Access Service Network, *see* ASN
Active mode, 227
Adaptive Antenna System, *see* AAS
Adaptive Modulation and Coding,
    see AMC
Admission control, 192
Advanced Encryption Standard, *see* AES
AES, 246–7
Aggregate bandwidth request, 138
AK, 210, 232, 234, 237–42
Alamouti, 200
Allocation Start Time, 123
AMC, 67
Application Service Provider, *see* ASP
Architecture, 207–17
    architecture reference point,
        see reference point
    network reference model, 209
    specifications, 207–8
ARQ, 106
    ARQ feedback, 106, 108

ASN, 209, 210–3
    ASN profiles 211–3
ASN gateway, *see* ASN-GW
ASN-GW, 210, 211
ASP, 210
Association, 220–2
ATM CS, 83
Authentication, 235–42
Authentication Key, *see* AK
Authentication, Authorization and
        Accounting, *see* AAA
Authorisation module, 87, 174
Automatic Repeat reQuest, *see* ARQ

Band AMC, 67
Bandwidth request header, 96
Bandwidth request opportunity size, 144,
    147, 149
Bandwidth stealing, 146
Base Station, *see* BS
Base Station ID, *see* BSID
Basic capabilities, 176
Basic connection, 102, 159
Beamforming, 195–9
BE, 164, 167
Best Effort, *see* BE
Bin, 67
Binary Phase Shift Keying,
    see BPSK
Block Turbo Codes, *see* BTC
Bluetooth 4, 6
BPSK, 45